Land Reform in Zimbabwe: Constraints and Prospects

Edited by

DR T.A.S. BOWYER-BOWER
School of Oriental and African Studies, University of London

DR COLIN STONEMAN
Centre for Developmental Studies, University of Leeds

Routledge
Taylor & Francis Group

LONDON AND NEW YORK

First published 2000 by Ashgate Publishing

Reissued 2018 by Routledge
2 Park Square, Milton Park, Abingdon, Oxon OX14 4RN
711 Third Avenue, New York, NY 10017, USA

Routledge is an imprint of the Taylor & Francis Group, an informa business

Publisher's Note
The publisher has gone to great lengths to ensure the quality of this reprint but points out that some imperfections in the original copies may be apparent.

Disclaimer
The publisher has made every effort to trace copyright holders and welcomes correspondence from those they have been unable to contact.

A Library of Congress record exists under LC control number: 00134809

ISBN 13: 978-1-138-74187-4 (hbk)
ISBN 13: 978-1-315-18263-6 (ebk)

Contents

List of Figures

List of Tables

List of Acronyms

AFC	Agricultural Finance Corporation (Zimbabwe)
AGRITEX	Department of Agricultural Technical and Extension Services (Zimbabwe)
ARDA	Agricultural and Rural Development Authority (Zimbabwe)
BZS	**Britain Zimbabwe Society**
CA	Communal Area
CAMPFIRE	Communal Areas Management Programme for Indigenous Resources
CAS	Centre of African Studies (University of London)
CFU	Commercial Farmers' Union (Zimbabwe)
CIDA	Canadian International Development Agency
DANIDA	Danish International Development Agency
DfID	Department for International Development (UK)
EIU	Economic Intelligence Unit (London)
FAO	Food and Agricultural Organisation
ICFU	Indigenous Commercial Farmers' Union (Zimbabwe)
ILO	International Labour Office
IMF	International Monetary Fund
IPFP	Inception Phase Framework Plan
LAMA	Legal Age of Majority Act (Zimbabwe)
LRP	Land Resettlement Programme (Zimbabwe)
LRRP-2	Land Reform and Resettlement Programme 2 (Zimbabwe)
LSCF	Large-Scale Commercial Farms (Zimbabwe)
LTC	Land Tenure Commission (Zimbabwe)
MAI	Multilateral Agreement on Investment
MLRRD	Ministry of Lands, Resettlement and Rural Development
NAD	Native Affairs Department (Rhodesia)
NFAZ	National Farmers' Association of Zimbabwe
NGO	Non Governmental Organisation
NLP	National Land Policy (Zimbabwe)
NLHA	Native Land Husbandry Act (Rhodesia)
NORAD	Norwegian Agency for International Development
ODA	Overseas Development Administration (UK)
OECD	Organisation for Economic Cooperation and Development

PASS	Poverty Assessment Study Survey (Zimbabwe)
RA	Resettlement Area
RO	Resettlement Officer
SOAS	School of Oriental and African Studies, University of London
SSCF	Small-Scale Commercial Farms (Zimbabwe)
UDI	Unilateral Declaration of Independence (Rhodesia)
UNICEF	United Nations Children's Fund
UNDP	United Nations Development Programme
WLSA	Women and Law in Southern Africa
ZANU-PF	Zimbabwe African National Union - Patriotic Front
ZAPU	Zimbabwe African Peoples' Union
ZFU	Zimbabwe Farmers' Union
ZHC	Zimbabwe High Commission
ZNFU	Zimbabwe National Farmers' Union

List of Contributors

Dr J. Alexander
Senior Associate Member, St Anthony's College, University of Oxford, Oxford, OX2 6JF, UK.

Dr T.A.S. Bowyer-Bower
Former Chairman, Centre of African Studies, University of London; and Lecturer in Geography, Department of Geography, School of Oriental and African Studies, University of London, Thornhaugh Street, London, WC1H 0XG, UK.

Jenny Brown
Formerly of the Institute of Development Studies, London School of Economics and Political Science, Houghton Street, London, WC2A 2AE, UK.

Professor Lionel Cliffe
Professor of Politics, Centre for Development Studies, University of Leeds, Leeds, LS2 9JT, UK.

Dr Simon Coldham
Senior Lecturer in African Law, School of Oriental and African Studies, University of London, Thornhaugh Street, London, WC1H 0XG, UK.

Professor John Cusworth
Professor of International Development Management; Dean of the Faculty of Social Sciences and Humanities; Head of the School of Social and International Studies and Director, Development and Project Planning Centre, University of Bradford, Bradford, West Yorkshire, BD7 1DP, UK.

Dr Jennifer A. Elliott
Senior Lecturer in Geography, School of the Environment, University of Brighton, Cockcroft, BN2 4AT, UK.

Dr Susie Jacobs
Senior Lecturer, Department of Sociology and Interdisciplinary Studies, Manchester Metropolitan University, Rosamund Street West, Manchester, M15 6LL, UK.

Dr B. H. Kinsey
Senior Research Fellow, Free University, Amsterdam, The Netherlands; and Senior Research Associate, Department of Rural and Urban Land Use Planning, University of Zimbabwe, P.O. Box MP 167, Mount Pleasant, Harare, Zimbabwe.

Professor Sam Moyo
Director of Studies, SAPES (Southern African Political and Economic Series) Trust, Zimbabwe; and Associate Professor of Agrarian Studies, Institute of Development Studies, University of Zimbabwe, P.O. Box MP 167, Mount Pleasant, Harare, Zimbabwe.

Dr Robin Palmer
Land Policy Adviser, Oxfam UK, 274 Banbury Road, Oxford, OX2 7DF.

Dr Debby Potts
Lecturer, Department of Geography, School of Oriental and African Studies, University of London, Thornhaugh Street, London, WC1H 0XG, UK.

Professor Mandivamba Rukuni
Programme Director, W.F. Kellogg Foundation Africa Programme; Research Professor of Agricultural Economics, University of Zimbabwe, P.O. Box MP 167, Mount Pleasant, Harare, Zimbabwe; and formerly Chair of the Land Tenure Commission of Zimbabwe.

Dr Colin Stoneman
Visiting Research Fellow, Centre for Development Studies, University of Leeds, Leeds, LS2 9JT, UK; and formerly Senior Research Fellow, Lecturer and Acting Director, Centre for Southern African Studies, University of York.

Acknowledgements

We are very grateful to all who helped in so many ways with the organisation of the March 1998 SOAS Conference, which was the precursor of this text. In particular we acknowledge the encouragement, enthusiasm and practical help provided by the Zimbabwe High Commission in London, the delegations from the government ministries and farmers' unions, and all speakers and other participants at the conference - both from the platform and the floor. The hard work of the Centre of African Studies (CAS), School of Oriental and African Studies (SOAS) of the University of London, and the Britain Zimbabwe Society (BZS), without which the conference would have been impossible, is also gratefully acknowledged. Sincere thanks are also extended to all those involved in the writing, editing and production of this text, and in particular, to Thomasina.

Disclaimer

Whilst all authors are responsible for the information they have included, no one author or editor of this text necessarily agrees with the views of other authors or editors.

1 Land Reform's Constraints and Prospects: Policies, Perspectives and Ideologies in Zimbabwe Today

DR T.A.S. BOWYER-BOWER AND DR COLIN STONEMAN

Background

Land reform directly or indirectly redistributes and/or redefines property rights to agricultural land (Carter, 1999). Land ownership, and rights of use of land, have been central issues for many countries throughout history, and for many are also issues of the day. They bear upon the livelihoods of rural and urban people, and can play a crucial role in economic development and environmental sustainability. Within this context, land reform has been pivotal to much development both throughout Africa and elsewhere (refer, for examples, to Dorner, 1992; Binswanger and Deininger, 1993; Putterman, 1995; and Chapter 2 of this text).

Land reform is often viewed in moral and political terms, as a necessary means by which land may be redistributed, for example to the landless and poor to help alleviate poverty, as reward for struggles for liberation, to help redress population-land imbalances, for example, brought about by apartheid regimes or unequal growth during colonial times, or as part of a package of agrarian reform aimed at boosting agricultural outputs. The need for land reform is often spurred on by slow industrial development and insufficient employment opportunities. Unless carried out for reasons of moral or political ideology alone, what is generally crucial is that land reform should not only address issues of equity, but also productivity. If the nation's economic development is also a consideration, it is important that land reform works to improve sustainable productivity from the precious land resource base of a country, and that the sustainability of national output as well as internal food security also be maintained if not enhanced, rather than threatened. How best to achieve this is often a topic of considerable contention and debate.

In Zimbabwe, for example, for the first ten years after independence in 1980, land redistribution was limited largely to that occurring on a 'willing-buyer, willing-seller' basis (as defined by the Lancaster House Agreement). During this time some 3 million hectares of commercial farmland was purchased by the government at market value, assisted by grants from the UK government (UK funds to be met 50:50 by the Government of Zimbabwe; including the1981 Land Resettlement Grant, which expired in 1996). Some 52,000 families, focusing on the landless and poor from overpopulated and environmentally degraded communal land, but also some war veterans, were resettled. There was considerable controversy surrounding the process and its management, as to whether the intended objectives were being fulfilled, the slow pace at which the reform was proceeding, and concerning the means by which the processes in place were enhancing or undermining economic and environmental development and stability. Many of these issues are addressed in subsequent chapters of this text.

With the expiry of the Lancaster House Constitution in 1990, the 1992 Land Act was passed, allowing for changes to procedure including, for example, compulsory land acquisition, and less certainty over what compensation would be paid for land acquired (Naldi, 1993). The Act also limited the size of farms, and introduced a land tax. Whilst aimed at pleasing those pushing for speedier progress of the land reform process, fears about the impact such action would have on the national productivity levels and the national economy were raised. There was also escalation in the argument about the efficacy of the reform schemes that had already taken place, argument over who was receiving the land (many of those receiving land were found to be President Mugabe's political associates and supporters; Binyon, 2000), and further debate on the process by which land reform should take place. In 1996 a policy paper on land redistribution and settlement was produced. However, in practice there was little resettlement of poor Zimbabweans in the early 1990s.

In November 1997 attempts were made by the government to accelerate the pace of reform by gazetting 1471 farms for immediate compulsory acquisition and with only partial compensation (some compensation being offered for infrastructural developments on the land but nothing for the land itself). This proposal was commonly termed 'land grab'. It was to procure a further 5 million hectares of commercial farmland for the proposed resettlement of an additional 110,000 families; and the contentions escalated accordingly.

With widespread support amongst all stakeholders (including those who stood to lose land) for the need for land reform in Zimbabwe to redress imbalance in access to the land resource base of the country which was otherwise unsustainable in the longer term (Moyo, 1995), it was the conditions by which this was to occur, which became the main cause for concern. Some felt the Government of Zimbabwe repeatedly used the land reform issue more as a rhetorical tool - as a populist vote-grabber, a threat to its critics, a reward for its favourites (*The Economist*, 1998; *The Daily Telegraph*, 2000); rather than as a means of actually making sound progress in helping the poor and landless, enhancing equity as well as productivity, without threatening the stability and sustainability of the land resource base of the country, and without damaging the national economy.

On 11 March 1998 an international conference was held at SOAS (School of Oriental and African Studies) in the University of London, entitled 'Land Reform in Zimbabwe: The Way Forward'. It was organised by the Centre of African Studies of the University of London (CAS) in conjunction with the Britain Zimbabwe Society (BZS) and the Zimbabwe High Commission (ZHC). It addressed the then controversial and highly publicised developments in the land reform issue, with the aim of seeking grounds for agreement amongst the diverse stakeholders on the way ahead. The conference drew on the expertise of top researchers in the field from around the world, as well as members of the Zimbabwe government responsible for the current Land Reform Programme, representatives of NGOs (non-governmental organization) involved in land resettlement in Zimbabwe and of funders involved in past resettlement, possible future funders, and representatives from Zimbabwe's three farmers' unions, including those who stood to lose land and those who stood to gain. This conference was the first open forum at which all parties came together for discussion and debate. Overall 22 speakers contributed to 13 presentations and a panel debate, at least half of whom were Zimbabweans, and 300 or so attended in the audience. A wide range of perspectives and research findings were aired and discussed in a good humoured and constructive atmosphere.

It is the presentations made during the conference, and debate from the plenary sessions, which have provided the main basis of this text. As such it is a unique statement on the very important and much discussed issue of land reform today. The text provides multi-disciplinary perspectives on this crucial issue, and provides a useful case study for informing comparative global, regional, and national research and understanding, as well as informing those involved in making and implementing policy.

Further chapters in this text

Chapter 2 (Palmer: *Mugabe's 'Land Grab' in Regional Perspective*) critically assesses Mugabe's recent 'land grab' approach to land reform in Zimbabwe, and calls for a more considered approach, drawing on strengths of contrasting attempts at land reform occurring concurrently elsewhere in east and southern Africa. Palmer's chapter is not only a useful overview of what progress is being made in land reform programmes in other parts of Africa (Kenya, Tanzania, Mozambique, Namibia, South Africa, Malawi, Zambia, Rwanda and Uganda being specifically mentioned), but also raises in regional perspective many issues that are later explored in more detail in further chapters of this text: notably the need for a range of flexible models of land reform, rather than one model being applied rigidly to all conditions prevailing (Chapter 16); implications of land reform for poverty (Chapter 8), rural welfare (Chapter 9), women (Chapters 14 and 15), and traditional rural-urban linkages (Chapter 10); implications for the land reform process of customary law and land tenure constraints (Chapters 13 and 14), and the possible value of a technicist approach (Chapter 11).

Chapter 3 (Cusworth: *A review of the UK ODA Evaluation of the Land Resettlement Programme in 1988 and the Land Appraisal Mission of 1996*) gives a critical overview of the land reform activities that took place in Zimbabwe between independence in 1980 and 1996 by presenting the main findings of the two official UK government reviews of the programme: 1988 and 1996. Useful insight is given into how the Land Reform Programme agreed at the Lancaster House talks was put into practice in the 1980s. The successes and problems encountered are analysed, and there is an attempt to explain the fading support - both from the government and donors - despite the programme's successes in poverty reduction and in its unusually high overall economic rate of return. Further to this, the main findings of the 1996 Land Appraisal Mission are presented, identifying what features the next phase of land reform should take for returns on investment and other benefits to be maximised. A main conclusion of this chapter is a warning for the process of land redistribution to be managed in an efficient and effective manner for the benefits to be maximised and sustained, which will require financial input from the international community, rather than allowing a lapse into a confidence sapping 'land grab' from which few can benefit.

Chapter 4 (Cliffe: *The Politics of Land Reform in Zimbabwe*) continues the theme of analysing the negative beliefs that arose about the land reform activities of the 1980s despite the overwhelming successes identified, and explores in detail motives behind perceptions and actions which have affected further progress in the Land Reform Programme. Whether there is the political will to fully commit the resources necessary to build up and maintain the institutional capacity required to fully implement further stages of the land reform process is questioned.

Chapter 5 (Stoneman: *Zimbabwe Land Policy and the Land Reform Programme*) summarises the Government of Zimbabwe's perspective on the land reform issues presented at the 1998 SOAS Conference. This includes an overview of the land issue, perspectives on the need for land reform, proposed objectives of the Land Reform Programme, analysis of the component features of the land reform process, and responses to concerns about implications of aspects of the programme that were aired at the conference.

Chapter 6 (Bowyer-Bower: *Theory into practice: Perspectives on Land Reform of the Farmers' Unions of Zimbabwe*) compares and contrasts views by Zimbabwe farmers' unions representing both those who stand to lose land and those who stand to gain land, on the land reform process. These views were amongst those put forward in presentations made by the Commercial Farmers' Union, the Indigenous Commercial Farmers' Union, and the Zimbabwe Farmers' Union at the SOAS conference. The distinctive emphasis of each union is highlighted, grounds for agreement are explored (in particular, the need for land reform, the need for a review of the land tenure system for land reform to be successful, views on who should be resettled, the need for transparency in how the programme of land reform is to be managed, the need for the programme to be undertaken within Zimbabwe's legal framework, views on the means by which land will be acquired), and further points of contrasting emphasis for each union is reviewed. Overall, how vital it is for all three unions to continue their positive collaboration in working together on the land reform issue is illustrated.

Chapter 7 (Moyo: *The Political Economy of Land Redistribution in the 1990s*) provides an insightful analysis of the 1471 farms gazetted by the government in November 1997 for compulsory acquisition. This includes analysis of ownership characteristics, gives insight to the politico-economic implications of land reform going ahead with regard to the various categories of farms included in the list. Good insight is also made into the political economy of the land reform process, and targets for optimising returns from land reform are suggested.

Chapter 8 (Bowyer-Bower: *Implications for Poverty of Land Reform in Zimbabwe: Insights from the Findings of the 1995 Poverty Assessment Survey Study*) explores what insight can be obtained from the findings of the poverty assessment survey undertaken in 1995 about the links between land reform and poverty in Zimbabwe. The results allow for comparisons to be made between rural land-use sectors (the large-scale commercial farms, small-scale commercial farms and resettlement areas, and communal areas) and urban areas, and interesting insight is obtained from an analysis of the incidence of poverty by land use sector, the incidence of poverty with land ownership, views of households experiencing poverty on the adequacy of land for grazing and cropping; and their perceptions of factors that identify poverty, the main causes of poverty, and solutions to poverty. The survey's findings emphasise the focus of householders on the need for the creation of employment and higher wages in urban areas as the main solutions to poverty, and identify the main needs for rural areas as being affordable agricultural loans and the provision of irrigation for overcoming the effects of drought and securing higher yields. A main message for the Land Reform Programme is the critical need for an adequate provision of support services for resettlement activities if vital goals of the Land Reform Programme are to be achieved.

Chapter 9 (Kinsey: *The Implications of Land Reform for Rural Welfare*) is an invaluable empirically-based evaluation of the successes of resettlement activities of the 1980s, by reporting the findings of ongoing field research (a longitudinal study) into the welfare implications of 400 families resettled in the first two years of the Land Reform Programme (1980-82) in Zimbabwe's three agriculturally most important agroclimatic zones. This is invaluable long term empirical research, which again throws into question much of the negative attitude towards whether the Land Reform Programme has been and can fulfil its stated aims. Indicators of welfare for the resettled families for 1997 are compared with those for 150 un-resettled households still living in communal areas. The considerably greater revenue, and other welfare benefits as defined by a variety of welfare indicators, obtained by the resettled families compared with those living in communal areas, are quantified and evaluated, and important implications of these findings for the Land Reform Programme are highlighted.

Chapter 10 (Potts: *Zimbabwean People's Perceptions of the Land Resettlement Programme: the Case of Rural-Urban Migrants*) reports on the views on land reform in Zimbabwe of 186 adults surveyed in 1994 who had recently migrated to low-income areas of Harare. This survey provides interesting insight into what the general views and points of understanding are amongst typical low-income urban Zimbabwe residents who otherwise contribute to no particular lobby group on the issues. The findings of the study highlight support for the programmes, and raise interesting points with regard to the value of resettlement to those resettled, the desire for the criteria of those eligible for resettlement to be broadened, and views on implications of land reform for the environment (explored further in Chapter 12). The survey also raises an interesting question about the extent to which resettlement activities should be directed, for example, to follow technical guidelines aimed at maximising productivity, or whether instead there is scope for their being an inalienable right to land, for which title deed should be granted, regardless of the use to which the land is subsequently put. This idea is explored further in Chapter 11.

Chapter 11 (Alexander: *The Enduring Appeal of 'Technical Development' in Zimbabwe's Agrarian Reform*) examines the history of technical direction being given to farmers to guide agricultural development in Zimbabwe. It explores technical direction being used as a political tool for coercion and control prior to independence, and the resistance to it by 'peasant' farmers when directed by authority under colonialism. This paper goes on to question the continuance of this 'technicist' approach to agricultural development since independence and, in the face of continued resistance to this approach today, questions the motives for, and validity of its use in agrarian reform. This view can be contrasted with the need for technical input suggested by the research presented in Chapter 12, recommendations suggested in Chapter 2 and Chapter 6, and the reported value of directed targets of behaviour and output mentioned by resettled women in Chapter 15.

Chapter 12 (Elliot: *Resource Implications of Land Resettlement in Zimbabwe: Insights from Woodland Changes*) points out the lack of any objective data or systematic monitoring of environmental impacts of land resettlement whilst a recurrent view in the literature is that land resettlement is likely to lead to a degraded environment all too typical of the communal areas that resettlement is meant to alleviate. The paper goes on to report the findings of empirical research based on aerial photograph interpretation and GIS analysis into the changes in resources presence (principally of grassland and woodland) in two resettlement areas subsequent to resettlement. Interesting insight is given into the various conditions of, perception of, and controls over resource use subsequent to resettlement. The continued importance of the woodland resources to the livelihood systems of resettled households is revealed, illustrating the importance of integrating woodland resources into the future planning of the Land Reform Programme if resource degradation is to be avoided. Also, a place-specificity of process and pattern in resource use is determined, indicating there can be no simple model of what resettlement practices do or do not contribute to sustainable resource management. The need for further research is indicated, and for flexibility in future models of resettlement if resource use is to be optimised.

Chapter 13 (Coldham: *Land Inheritance Issues in Zimbabwe Today*) examines shortcomings in the rules and practices governing land inheritance in Zimbabwe that were revealed in the findings of the 1993 Commission of Enquiry into appropriate agricultural land tenure systems, under the chairmanship of Professor Rukuni, as having a significant impact on the likely effects of the Land Reform Programme. The likely effects of land inheritance law for large scale commercial farm (LSCF) areas, small scale commercial farm (SSCF) areas, resettlement areas (RA) and communal areas (CA) are explored. The case is put for abolishing customary land tenure as, for example, is prevalent in the communal areas, and replacing it with freehold/leasehold title systems currently operating in LSCF areas and SSCF areas. The need for freehold/leasehold title systems in resettlement areas is also illustrated. This important implication of land tenure for the sustainability of benefits of land resettlement is also raised in Chapter 6 and Chapter 16 of this text.

Chapter 14 (Brown: *Land Reform versus Customary Law: What About Women?*) looks at gender inequality in land issues in Zimbabwe, despite recent changes in legislation intending to give women a legal status equal to men. The history of women's rights with respect to land in Zimbabwe are reviewed. The gap that is evident between recently legislated equality, and the still practiced Shona customary law which excludes women from independent rights to land and maintains them in a state of perpetual legal incapacity, is explored. Changes that would help achieve greater gender equality in practice, and the advantages this would bring to the sustainability of intended benefits of the Land Reform Programme are illustrated.

Chapter 15 (Jacobs: *The Effects of Land Reform on Gender Relations in Zimbabwe*) reports on the findings of field research undertaken 1983-84 into implications of resettlement for women resettled 1980-82. The findings of the research reveal implications for aspects of their wealth and status individually, within the family, community and within society. The research determines that despite women's situation of *de facto* legal, economic and structural dependence, resettlement has nevertheless benefited wives and widows in several respects. Negative implications of land inheritance rights and land tenure (also explored in Chapters 13 and 14 of this text) are again illustrated. Interesting insight is also given into the knock-on effects of land resettlement for the structure and functioning of the family unit and for gender relations.

Chapter 16 (Rukuni: *Land Reform in Zimbabwe - Dimensions of a Reformed Land Structure*) puts into context the land reform that has taken place in Zimbabwe so far, and suggests a 'national strategy' for the future Land Reform Programme of the country. This is based on stated government intentions, enriched with positive elements from views of the Commercial Farmers Union, as well as points and issues raised by the Zimbabwe Farmers' Union, the Indigenous Commercial Farmers Union, and other key stakeholders in the private sector and civil society. Political stability, establishing a broader basis for economic growth, and the need for social integration, are the objectives focused upon. Values and principles for inclusion in the national strategy, many of which have been the focus of other chapters of this text, are presented (for example, the need for greater security of tenure, efficiency of land use, effective farmer support). The need to recommence the Land Reform Programme without major short-term political shocks and economic instability is considered crucial, and a win-win solution to current impasses is sought. Rukuni goes on to propose four new categories of land reform models, elaborating new resettlement/reform options in each that will broaden access and provide greater flexibility than has existed so far, and that has been called for by a number of other chapters of this text. Proposed important components of the necessary institutional and legal framework and programme management structure are presented, and plans for financing the scheme are considered. It is concluded that a number of fundamental and structural issues of the current Land Reform Programme need to be addressed to avert a future crisis in national prosperity and social progress if instead the land issue remains unresolved. The need for the new programme to be backed by all key players including the private sector and civil society is emphasised.

The Way Forward

Two years after the 1998 SOAS conference very little progress has been made towards a resolution of the issue, despite increasing domestic and international concern. Many of the 1471 farms gazetted for immediate compulsory acquisition in November 1997 were de-listed on appeal whilst others were offered for sale to the government for agreed compensation to be paid. In September 1998 a Donor's Conference was held in Zimbabwe. This set out principles for a second phase in Zimbabwe's Land Reform and Resettlement Programme (LRRP-2) that were agreed by all parties (including government, international donors, commercial farmers, the private sector): transparency, support for the law, poverty reduction, affordability and consistency with Zimbabwe's wider interests. This would include proper compensation being paid for land acquired. The principles under which LRRP-2 would commence with a two-year inception phase initially using 118 farms being offered (about 200,000 ha) for compulsory acquisition, were agreed, along with a total area target of about 1 million ha for the inception phase (van den Brink, 2000). The Inception Phase Framework Plan (IPFP) detailing how this programme would be developed and implemented was subsequently produced (Government of Zimbabwe, May 1999), and donors set about agreeing to provide financial support for the programme.

In November 1999, for example, the World Bank and Zimbabwe signed an agreement for a US$5 million loan to support land reform, with the World Bank's country director for Zimbabwe stating:

> The Bank fully supports the government's Inception Phase policy framework and will now start to finance the actual resettlement of poor farmers with this credit. I believe that successful implementation of land reform is key to resolving one of the most fundamental equity and efficiency issues facing the country as it enters the new millennium (World Bank, 1999).

In reply, Senior Minister Joseph Msika, who was responsible for the Land Reform Programme said:

> I am very pleased that the World Bank has taken the lead on fulfilling the pledges made at the 1998 Donor Conference. I hope and trust that many other donors, stakeholders and NGOs will follow (World Bank, 1999).

President Mugabe, however, continued to make public statements and instigated and encouraged actions that were at variance with the position agreed at the 1998 Donor's Conference (van den Brink, 2000). For example, in late January 2000 President Mugabe amended Clause 57 of the draft constitution to read:

> The former colonial power has an obligation to pay compensation for agricultural land compulsorily acquired for resettlement through a fund established for the purpose. If the former colonial power fails to pay compensation through such a fund, the Government of Zimbabwe has no obligation to pay compensation for agricultural land compulsorily acquired for resettlement.

Britain stated that no country could unilaterally impose an obligation on another. The Government of Zimbabwe's argument rests on reassurances given in the late 1970s in the negotiations to broker an end to the liberation war that was ended with the 1979 Lancaster House Agreement giving rise to independence in 1980. Land was a key issue throughout these negotiations, and general assurances were given that the international community would assist with this problem (Palmer, 1990; Holman, 2000). As stated at the beginning of this chapter, the UK government has assisted with the first phase of the Land Reform Programme instigated at independence, and Britain along with other donors have pledged further assistance if agreed conditions are adhered to. It may well be that total international assistance in the past has fallen well short of the level of assistance that might have been hoped for (Anderson, 2000). Anger at this outcome, however, would have been more appropriate in the early 1980s. Since then nearly two decades have gone by during which the Government of Zimbabwe has under-spent even the sums offered by Britain, has allowed the programme of land reform to nearly wither away, and land acquired for resettlement has been received by President Mugabe's political associates and supporters (Binyon, 2000). In this light the current anger of the Government of Zimbabwe thus seems somewhat artificial.

When the referendum on the draft constitution was held on 20 January 2000, the Zimbabwean electorate rejected the constitutional changes as putting too much power in the President's hands (Dorman, 2000). It thus appears that if the Clause 57 had been intended to 'buy' the electorate, as earlier promises of land reform close to election times appear to have done, its authors under-estimated the growing sophistication (or cynicism) of Zimbabwean voters.

Undaunted, the President announced in early March (2000) that an amendment was to be made to the existing constitution along the same lines, and this was duly passed by Parliament on 6 April. With the confiscation of land without compensation now made 'legal' by an Act of Parliament, the scene was set for farm occupations by so-called 'veterans' of the liberation war that had started in February and reached a peak in April 2000 with about a thousand farms being occupied, sometimes violently, with some loss of life.

The Zimbabwean government claims that these land invasions are nothing more than a spontaneous desire by the landless to redress historical grievances and take back land that was taken from them without compensation. It is widely accepted, however, that these occupations and the violence are more related to an attempt by President Mugabe and the ruling party (ZANU-PF) to avoid a likely defeat in the general election due by mid-2000. Farmers who have allowed political opposition party candidates to address their workers have been particularly targeted, and some murdered, and a condition for other farmers to return to their land has been that they would denounce the opposition. Farm workers, and more recently, schoolteachers, have also been a particular target of the violence. Meanwhile, most of the so-called 'war veterans' are ZANU-PF supporters, often far too young to have fought in the war, and being paid for their actions.

There are many casualties of a strategy such as this. One is any viable poverty-focused land-reform programme, for there are no domestic funds available to develop any confiscated lands and international economic support has been frozen as the agreed conditions upon which funding would be made available have been broken. It is feared that the wider economy may be another. Questions are now being asked about what the likely implications of these actions are for land reform programmes elsewhere in Africa (Cousins, 2000). The insecurity in the farming regions is severely hampering the harvesting and processing of this year's crops which contribute a third of exports. Preparation of farming regions for next year's crops is severely hampered. Investment confidence has also undoubtedly been severely shaken.

So this book ends with an impasse: all agree that land reform is essential, but disagreement among the parties on how it should be undertaken has made sustainable land reform impossible. After the general election, it is hoped that it will be possible for a new start to be made, and that the considerable goodwill shown in the past by all parties towards finding a balanced outcome for the nation's land question can be rekindled. There are undoubtedly many lessons to be learnt from the past both by the government and the international community. Difficult and painful though these may be, only once they have been taken to heart is it likely that future cooperation will succeed in finding a sustainable solution that addresses a satisfactory balance of the objectives of land reform that is acceptable to all parties. It will take real courage from all sides for the impasse to be broken. Evidence suggests, however, that the consequences of not doing so are an even more painful and considerable loss for all.

[A postscript is placed at the end of this book.]

References

Anderson, D. (2000), 'Mugabe is right about Land Reform' *The Independent*, 4 May, 2000.

Binswanger, H. and Deininger, K. (1993), 'South African Land Policy: Historical Legacy and Current Policy Options', *World Development*, 21, 9.

Binyon, M. (2000), 'How Mugabe Abused Backing from Britain', *The Times*, 19 April.

van den Brink, R. (2000), *Zimbabwe Land Reform Update*, 2 March, 2000, World Bank, Zimbabwe.

Carter, M.R. (1999), 'Land Reform', in P.A. Hara, *The Encyclopaedia of Political Economy*, Routledge, London.

Cousins, B. (2000), 'Why land invasions will happen here too', article submitted to the *South African Mail and Guardian*, 28 April, 2000.

The Daily Telegraph (2000), 'Mugabe gives white farms to his cronies: British fury as officials grab 1 million acres', *The Daily Telegraph*, 29 March, 2000.

Dorman, S.R. (2000), 'Change Now', *The World Today*, Vol. 56, 4.

Dorner, P. (1992), *Latin American Land Reforms: A Retrospective Analysis*, University of Wisconsin Press, Madison.

The Economist (1998), 'Populism Awry,' *The Economist*, 24 January, 1998.

Government of Zimbabwe (1999), *Inception Phase Framework Plan (IPFP) 1999 - 2000: An Implementation Plan of the Land Reform and Resettlement Programme Phase 2*, prepared by the Technical Committee of the Inter-Ministerial Committee on Resettlement and Rural Development (IMCRD) and the National Economic Consultative Forum's Land Reform Task Force, Government Printers, Harare.

Holman, M. (2000), 'A shaky grip on Zimbabwe's moral high ground', *Financial Times*, 13 April, 2000.

Moyo, S. (1994), *Economic Nationalism and Land Reform in Zimbabwe*, Occasional Paper No. 7, SAPES Trust, Harare.

Naldi, G.J. (1993), 'Land Reform in Zimbabwe: Some Legal Aspects', *Journal of Southern African Studies*; 31.

Palmer, R. (1990), 'Land reform in Zimbabwe, 1980 – 1990', *African Affairs*, 98.

Putterman, L. (1995), *Continuity and Change in Rural China*, Cambridge University Press, Cambridge.

World Bank (1999), *World Bank News Release No. 2000/099/AFR*, 16 November, 1999.

2 Mugabe's 'Land Grab' in Regional Perspective

DR ROBIN PALMER

Introduction

Robert Mugabe's attempted land grab has been widely criticised at home and abroad. It has generally been perceived as a crude attempt to deflect attention away from growing opposition and mounting, often self-inflicted, economic problems by finding a convenient and easy scapegoat. Ageing presidents, approaching 20 years in office, surrounded only by praise-singers and out of touch with domestic and international realities, have already brought much economic and social damage to Zambia, Malawi and Kenya. The same trend has now become apparent in Zimbabwe. A deeply corrupt and widely unpopular regime today stands indicted within and without, even by those who genuinely wish Zimbabwe well and were once firm supporters of its government. When friends of Zimbabwe find themselves nodding in agreement with Ian Smith and the International Monetary Fund (IMF), we have indeed entered strange waters.

The specific criticisms of Mugabe's land grab can briefly be summarised: lack of funds, lack of planning, lack of capacity, lack of accountability and, at the Commonwealth meeting in Edinburgh, spectacular lack of diplomacy; a recent history of rewarding its own supporters more than landless peasants; a list of 1,480 farms gazetted for compulsory acquisition so full of errors, so far removed from agreed past criteria, and so politically motivated as to be a subject of derision, had it not proved economically damaging both within Zimbabwe and to Zimbabwe's image abroad. The whole affair must be deeply embarrassing to those within government who are attuned to current economic realities, but are unable to prevent their president from continuing to posture. Though internal struggles are clearly continuing, producing very mixed messages, it is probable, as the Economic Intelligence Unit (EIU, 1998) suggests, that something akin to the Commercial Farmers' Union's (CFU) *Team Zimbabwe* concept represents the best prospect of modest but effective land reform, especially if conceived on a largely self-financing basis. What this paper will focus on, however, is the regional perspective, in particular, the possible impact of Mugabe's land grab on prospects for land reform elsewhere in southern and east Africa.

Land Reform in Context

It matters when countries achieved independence or majority rule. The economic choices open to Tanzania in the 1960s, to Mozambique in the 1970s, to Zimbabwe in the 1980s, and to South Africa in the 1990s were all distinctively different. South Africa, for example, is currently engaged in a three-pronged programme of land tenure reform, redistribution and restitution in a far more constraining environment than that which initially confronted Zimbabwe. The dominant 'new paradigm' now on offer in South Africa (as in Brazil and Colombia) is one of market-assisted land reform. Under this, the state's role is limited to providing financial support to individuals, but more usually groups, trying to buy land from commercial farmers on the famous 'willing-buyer, willing-seller' basis. Market forces are expected to iron out any difficulties. One of the problems with this approach - and there is an obvious similarity with Zimbabwe in 1980 - is that it is ahistorical. It effectively ignores all that has gone before (which is very widely known) and it also ignores the current reality that power on the ground still resides very much with the white commercial farmers (described in South Africa as 'organised agriculture') who are in a position to dictate terms - and prices - to would-be buyers, and whose lack of enthusiasm for redistribution is well known.

Current problems in South Africa's land reform programme in part mirror earlier Zimbabwean experience. White farmers have evicted thousands of people in anticipation of legislation designed to afford tenants and farm workers greater protection. They remain powerful, organised and fundamentally opposed to land reform. The Department of Land Affairs seriously lacks the capacity to implement land reform at both national and provincial level. Some of its officials from campaigning NGO (non-government organisations) backgrounds are wedded to communal or collective approaches which are not always most appropriate. The old Ministry of Agriculture remains largely unreconstructed and totally technicist. The wheels of the courts investigating historical claims to land grind exceedingly slowly. The land reform pilots are the land reform programme. There is growing impatience at the slow pace of land reform and a warning from Derek Hanekom, who holds the difficult combined post of Minister of Lands and of Agriculture, that unless it is speeded up, he may face increasing pressures to go for a Mugabe-style land grab.

There is currently a great deal of serious conflict over land throughout southern and east Africa (Palmer, 1997; Matowanyika and Marongwe, 1998), but none of it has attracted anything like the publicity or attention of recent events in Zimbabwe. There are, for example, controversial and highly contested land bills due before parliaments in Uganda and Tanzania later this year (1998). The Uganda Land Alliance and the Tanzanian National Land Forum have been leading campaigns to protect community land rights, which they see as seriously threatened by these bills. An important land law (Lei de Terras) was passed in Mozambique in July 1997 which, to the surprise of many, gave peasants significant new rights, and the NGOs ORAM (Associação Rural de Ajuda Mútua) and UNAC (União Nacional de Camponêses) are now involved in a campaign designed to enable peasants to assert those new rights. A land commission with a very wide remit has been sitting in Malawi, following earlier ones in Zimbabwe, Mozambique, and Tanzania. Activist NGOs have formed land alliances or coalitions in Uganda, Zambia and Tanzania, as earlier in South Africa, and new coalitions may be in the process of formation elsewhere. They are pressing hard for governments to allow meaningful national consultation and debate before land laws are passed, recognising government tendencies to work in secret and to exploit poor people's lack of information and power. Some also lobby donors such as the World Bank and the UK Department for International Development (DfID), who remain actively engaged in land issues. These days donors tend to stress (as recently too, Mugabe) the need for accountability, transparency, and a poverty focus.

Donors also in principle support NGO demands for full consultation, though who should do this and how, and who should be consulted, are always contentious political issues. The absence of an NGO land alliance in Zimbabwe can perhaps be put down to a more differentiated society and hence competing class and race interests over land, some of which were represented at the SOAS (School of Oriental and African Studies) conference.

Land Reform and the International Economy

The context behind all this conflict over land is complex. At the risk of huge oversimplification it is the impact which current economic orthodoxy - and the emphasis on privatisation and market forces in particular - has on access to land, which is causing so many problems, especially coming as it does after half a century and more of state interventions in the economy. Governments in Africa now find themselves under great pressure - and competing with each other - to open up to foreign investors in what, in an era of globalisation, is very much an investors' market. This can involve 'selling off the family silver', as Harold Macmillan so memorably characterised Margaret Thatcher's privatisation programme in Britain. In Africa, the family silver has come to mean minerals, land, and even water.

Within the context of current land struggles it is possible to distinguish three different categories of countries in southern and east Africa. The first comprises South Africa, Zimbabwe and Kenya, which have all known substantial alienation to white farmers in the past and have sought the very difficult and complex task of seeking to redress their colonial heritage. Here resettlement, in one form or another, of black farmers onto formerly white-owned land is seen as a key priority, though one of the peculiarities of land reform discourse in Zimbabwe is that it is so often couched exclusively in terms of resettlement. There is also a curious fixation with the 'magic' number of 162,000 families needing resettlement. In fact this derives from an original Muzorewa era estimate of 18,000, which was multiplied by 3 around the time of Independence to give a total of 54,000, then multiplied again by 3 to give 162,000 a few years later. Though this figure has absolutely no basis in reality, it continues to figure prominently in the minds of planners and politicians and is religiously repeated in government pronouncements on land reform.

The second category includes Tanzania, Uganda and Zambia, which saw very little colonial white settlement, but which seem now almost indecently anxious to compensate for that lack by throwing themselves open to what virtually amounts to a second scramble for Africa. The third includes countries like Malawi, Namibia and Mozambique, which witnessed alienation only in certain parts and so tend to. exhibit both tendencies. Southern Malawi also shares with Rwanda the disturbing Rift Valley phenomenon of acute landlessness (André & Platteau 1996) and related tendency for coping mechanisms to cease coping, and for viable options to be depressingly few (ELUS, 1997).

So in Tanzania, Uganda, and Zambia there has been, and continues to be, great pressure to 'open up' and pass legislation to make land easily available to new investors, whether local, foreign, or a combination of the two. Such pressures could well increase if the OECD's (Organisation for Economic Cooperation and Development) controversial Multilateral Agreement on Investment (MAI) comes into force. Opening up applies especially to tourist ventures, in which most countries in east and southern Africa find themselves competing with each other. Game parks, theme parks and private lodges for the seriously rich have sprung up all over the place, with adverse effects on pastoralist communities in particular, as huge areas of once common grazing land have been fenced off. Increasing mining activities in east Africa are also directly threatening land rights in places like Karamoja, northern Uganda. Internally there are strong pressures from those who wield political or economic power to turn this situation to their advantage. In Uganda, for example, landlords have been pressing for the end of remaining state controls in order to be able to exploit 'their' tenants without restraint. In fact at the end of heated debates on the land bill, landlords in Buganda lost out to tenants. But the conflicts are unlikely to go away. More generally, deals are usually struck in an atmosphere of corruption and secrecy, so that local communities are often the last to know that their land has been signed away - as was the case of course in the days of Rhodes and Lobengula.

Individual Title, Top-Down Approaches and Devolution

In addition to such pressures, modernisers in governments often genuinely believe (like their colonial predecessors) in a variant of the survival of the fittest, in which individual title to land is seen as the only way forward, while all indigenous forms of tenure are rubbished as regressive or worse, and landlessness is just one of the social costs you have to pay for progress (or someone else has to pay). In many African officials' minds the Kenyan titling and registration model is still seen as the solution, even though the World Bank itself no longer promotes it as once it did and the 'verligte' wing of the Bank has openly admitted its many failings, acknowledging that titles in Kenya have become 'virtually worthless' because landowners have no incentive to update them (Deininger and Migot-Adholla, 1997).

A persistent theme throughout land reform and resettlement programmes has been a top-down, directive, controlling approach which assumes that officials know best and that peasants and pastoralists need to be told what is best for them. Zimbabwe's resettlement programme, often unfairly criticised, certainly falls into this category. The resettlement areas have been consciously conceived and run as a separate world from the communal areas with no attempt to explore more flexible models. This may change should the models of Rukuni (Chapter 16 of this text) be considered in future shapings of Zimbabwe's Land Reform Programme. South Africa is trying to avoid such rigidity by allowing for a variety of different forms of tenure, but practice on the ground also tends to be rigid. In addition, inflexible thinking has allowed many settler myths about African agriculture to survive long after they should have been buried (Pankhurst, 1996). Too much attention is paid to the rationality of planners, too little to that of peasants and pastoralists, while the voices of peasants are almost never heard at conferences and workshops.

A key issue in current land reform debates is whether power over allocation, management and sale of land (and other natural resources) should remain centralised or be devolved to local structures. In other words, the issue is whether power should reside with the ministry or the village, and then, with whom in the village. This is central to the current debate in Tanzania (Shivji, 1998). The role of traditional authorities in this sphere has also had an interesting and somewhat cyclical history in Zimbabwe and elsewhere. Reviewing the region as a whole, Clement Ng'ong'ola has suggested that the decentralised Botswana land board system has provided an adaptable legal framework for often intractable problems of customary land tenure reform, one which can accommodate evolutionary or revolutionary processes, and could usefully serve as a model elsewhere (Ng'ong'ola, 1996). Refer also to Chapters 13 and 14 of this text for discussion of these points with regard to Zimbabwe.

Land Benefits versus other Economic Opportunities

But do people really want land in Zimbabwe or South Africa, and if so, for what purposes? In December 1997, the *Electronic Mail and Guardian* published a piece headed 'The people don't even want Mugabe's land'. It cited a study on poverty by the Ministry of Welfare (explored further in Chapter 8 of this text), which had asked 18,000 rural and urban households what were the main causes of poverty and how they could be combated. It found that people 'want jobs in a market economy, and an opportunity to work for a decent living.' Only one per cent of those responding said poverty was caused by a shortage of land, only two per cent that poverty could be resolved by the provision of land. Those who did want land, the article continued, included the elite, ex-combatants, and indigenous black commercial farmers (Electronic Mail & Guardian, 1997). A collection of interviews in the same paper in March 1998 carried an interesting range of views. For the unemployed Willard Mkuwasenga: 'Among my friends, I am the only one keen on a plot of land - if provided with credit, tools and seeds. Most of my friends prefer jobs to land. But they were born in the city. I come from the rural areas.' For the handicrafts seller Gertrude Gaza, 'Forty-five years in Mabvuku (Harare) have not made me an urban woman. Harare is a transit point. You must have a home in the rural areas. When you are retrenched, unemployed or sick, you go back to the rural areas. Where the ancestors are buried. I want a small commercial farm in Bindura.' (Electronic Mail and Guardian, 1998). These responses crystallise the complexity of the subject in urbanised countries with long histories of labour migration. There certainly are demands for land from different groups (including women wanting land in their own right) and for different purposes. What planners often assume wrongly is that people want land for full-time agricultural purposes, whereas in the majority of cases they want it as part of a range of survival strategies, to try to stabilise income when other sources of income are erratic, as they increasingly are in an era of structural adjustment and diminishing urban employment. Refer to chapters 8 and 10 of this text for further debate on these issues.

This raises the question of whether the planners of the 1951 Native Land Husbandry Act might have got it right after all. Famously described by George Nyandoro as 'the best recruiter Congress ever had' and the instigator of much rural insecurity and unrest, the Act attempted to break the by now traditional links between urban and rural and to turn Zimbabweans into either full-time workers or farmers (a theme explored further in Chapter 11 of this text). Today's optimistic modernisers argue that this indeed has to happen, that Africa's future lies on the path to industrialisation and that people will have to leave the land. An extreme version of this was voiced by Charles Onyango-Obbo, Editor of The Monitor of Uganda: 'It sounds heartless, but behind closed doors, businessmen and the smarter politicians will tell you that the solution to the hunger in Africa is to chase the peasants off the land. Not by force, mind you, but through subtle market mechanisms and simply buying them out' (The East African, 1997). The land NGOs currently struggling to defend peasant and pastoralist land rights do not share that vision of Africa's future, which they see as highly unlikely under globalisation. They argue instead that anti-poverty strategies can only succeed on the basis of economic growth and equitable distributions of incomes and that to raise incomes people need more, not less, secure access to land in order to participate in economic growth, as happened to a degree in South Korea and Taiwan.

A Grim Outlook

Mugabe's land grab is irresponsible, in the regional as well as in the national context, and in the last analysis because it gives ammunition to enemies of the poor. It can be of no benefit or comfort whatever to groups like the Uganda Land Alliance, which are struggling to defend previously secure community land rights in the current investor-friendly climate and within the current rules of the game. They know that to achieve anything they have to operate within those rules, whether they like them or not. So they engage with and lobby relevant donors precisely because it is they who are able to influence and pressurise governments, or to 'arm-twist' as Mugabe said of the International Monetary Fund (IMF). The Alliance has also engaged with the Ugandan Ministry of Lands, forcing it over time to take their concerns and arguments seriously. They now run workshops jointly with the ministry as part of the process of consultation. They and other non-government organisations effectively targeted members of Parliament in a workshop and public meeting in Kampala in March 1998 just before the final draft of the land bill was put before parliament. They have come to learn from experience that there can be an effective *de facto* working alliance of the powerless and the powerful.

The kind of dialogue taking place in Uganda, difficult and problematic though it has been at times, is clearly what needs to be restored in Zimbabwe, as many Zimbabweans are only too keenly aware. For if Mugabe is not persuaded to return to the paths of negotiation and transparency, the economic future for Zimbabwe may well be very grim indeed.

References

André, C. and Platteau, J. P. (1996), 'Land Relations under unbearable Stress: Rwanda caught in the Malthusian Trap', *Seminar on Land Tenure and Tenurial Reform in Africa*, London School of Economics, May 1996.

AWEPA (European Parliamentarians for Southern Africa) (1997), *Mozambique Peace Process Bulletin*, 19, September 1997.

Deininger, K. and Migot-Adholla, S. (1997), 'Principles and Evolution of the [World] Bank's Land Policy: Implications for the Ugandan Draft Land Bill', *Land Tenure seminar*, Kampala, 2 September 1997.

The East African (1997), 29 September - 5 October.

Economist Intelligence Unit (1998), *Zimbabwe Country Report, 1st quarter*, EIU, London.

Electronic Mail and Guardian (1997), People don't even want Mugabe's land', 1 December.

Electronic Mail and Guardian (1998), 'This land is your land, this land is our land', 2 March 1998.

ELUS (1996/7), *Malawi Estate Land Utilisation Study*, Government Printers, Malawi.

Matowanyika, J.Z.Z. and Marongwe, N. (1998), *Land and Sustainable Development in Southern Africa: An Exploration of Some Emerging Issues*. ZERO Sustainable Land Management Working/Discussion Paper Series 1, Harare.

National Land Forum (1997), *Azimio La Uhai: Declaration of NGOs and Interested Persons on Land*, Hakiardhi, Dar es Salaam.

Ng'ong'ola, C. (1996), 'Customary Law, Land Tenure and Policy in some African Countries at the Threshold of the Twenty-first Century', in G.E. van Maanen and A.J. van der Walt (eds), *Property Law on the Threshold of the 21st Century*, Maklu Press, Antwerp.

Palmer, R. (1997), *Contested Lands in Southern and Eastern Africa: A Literature Survey* Oxfam Working Paper, Oxford.

Pankhurst, D. (1996), 'Unravelling the Myths: Land Tenure, Agriculture and Land Reform in Independent Namibia', *Seminar on Land Tenure and Tenurial Reform in Africa*, London School of Economics, May 1996.

Shivji, I. (1998), *Not Yet Democracy: Reforming Land Tenure in Tanzania*, International Institute for Environment and Development and Hakiardhi, London and Dar es Salaam.

Uganda Land Alliance (1997), *Open Letter to the Minister of Lands, Housing and Physical Planning on the proposed Land Bill of 1997*. Uganda Land Alliance, Kampala.

The Zimbabwe Independent (1997), 7-13 November, 1997.

3 A Review of the UK ODA Evaluation of the Land Resettlement Programme in 1988 and the Land Appraisal Mission of 1996

PROFESSOR JOHN CUSWORTH

Introduction

This chapter summarises and reviews two studies instigated by the UK Overseas Development Administration (the ODA, now Department for International Development - DfID) into aspects of land resettlement in Zimbabwe. The first, in 1988, was a preliminary evaluation of the first Land Reform Programme (LRP) implemented with substantial financial assistance from the UK and other donors since independence in 1980. The second was a report on a land appraisal mission aimed at 'drawing up proposals for a new Land Reform Programme to be assisted by the British Government' undertaken eight years later in 1996. Given that the findings of both reports were generally positive about land resettlement, in terms of its ability to deliver economic and social benefits to poorest sections of the population, the question arises as to why UK and other donor support for resettlement faded away between the late 1980s and mid 1990s. Some possible explanations for this are raised in the chapter.

Background

Land resettlement was a key issue at the Lancaster House talks which led to independence in 1980. The UK ultimately pledged £20 million towards an initial programme to resettle 17,500 families on 1.1 million hectares of commercial farm land. A further £10 million was made available in 1981, in order to assist the Zimbabwe government in meeting its commitment of funding the programme on a 50:50 basis. In addition to financial assistance, the UK played a major advisory role in programme implementation through the procedures established to plan, appraise and monitor the implementation of individual schemes. Whilst the ODA insisted that the programme should aim to maximise economic benefits in a sustainable way it did not have a central role in the overall design and detail planning of the programme, which was undertaken by local ministries and departments.

In 1984, at the request of the Zimbabwe government, the ODA provided technical assistance for the establishment of a monitoring and evaluation unit (M&E Unit) within the ministry responsible for programme implementation. The M&E Unit set up an annual sample household survey system in 1985 across a range of Model A schemes which collected data on farm incomes and other socio-economic indicators. In 1988 the ODA commissioned a preliminary evaluation study of the programme. This study drew heavily on the M&E Unit data and other studies carried out by the unit. The final report of this study (Cusworth and Walker, 1988) was published by ODA in September 1988.

The following section of this chapter summarises the findings of this evaluation study and reviews the more controversial issues which it raised.

The 1988 Evaluation Report - Summary Findings

The first major finding of the study was simply one of observation, i.e. that by the end of 1987, just seven years after Independence, the Land Reform Programme had involved the resettlement of approximately 40,000 households on over 2.2 million hectares of land at the cost of about £80 million. The fact that most of this resettlement was undertaken in a well organised and planned manner associated with heavy investment in social infrastructure, indicated that this was an impressive achievement. In the wider context the report concluded that 'in particular the programme undoubtedly achieved its short run political objective of contributing to post war reconstruction and stability'.

The second major conclusion of the evaluation was that the majority of settler households had benefited considerably under the LRP in at least one of two ways either through increased opportunity for income generation and/or improved access to water, health and education services. Given that most resettled families were drawn from the most deprived sector of the population the report concluded that the LRP therefore had a positive impact on poverty reduction.

The third generally positive finding of the report was perhaps the most controversial. This was that, on the basis of the data taken from the M&E survey of Model A schemes (individual household based schemes) over several years, and against a number of assumptions about the Zimbabwe economy, the economic rate of return on the LRP from the perspective of the national economy was very positive and estimated at 21 per cent. This is a very high figure for any development project or programme. This conclusion proved to be particularly controversial at the time of publication in late 1988 when the popular perception of the LRP was that is was 'failing'.

The methodology employed for determining the economic worth of the LRP during the study is, however, defensible. This was a standard economic cost-benefit analysis using a range of assumptions about various economic variables. An important one was the opportunity cost of much of the land resettled under the early part of the programme. This was considered to be very low as it was mostly previously abandoned or used for extensive cattle grazing. Under resettlement a considerable proportion of this land was cultivated thus yielding a higher economic output per unit area. Another important set of assumptions in the analysis involved making adjustments to the financial value of farm output and inputs to reflect their economic value by valuing them at border parity prices using a further adjustment to reflect the relative scarcity of foreign exchange at the time. It might be argued that the empirical data used were inaccurate but, even accounting for a high margin of error, the overall result appears to endorse the finding that by putting relatively under-utilised land previously used for capital intensive, extensive cattle grazing under more intensively used arable farming using low value inputs, mostly labour, the rate of return is very high. It can still be argued that the economic benefits from converting relatively under-utilised commercial farm land, under highly capital intensive farming, to intensive small-holder production, using low-capital, high-labour husbandry techniques, are likely to be positive.

However, despite these positive findings the evaluation identified a range of problem areas affecting the LRP, some of which posed a real threat to the sustainability of the economic benefits.

One problem identified was that the benefits of the LRP were highly skewed across the settler population. There were indications that many settler households appeared to have been unable to benefit from the programme. Various explanations for this were given including lack of ownership of draught cattle, indebtedness, insecurity and gender-related problems. These problems had resulted in a high proportion of settlers being unable to make productive use of the assets allocated to them under the programme. Added to this some resettlement schemes were established in areas where, due to the natural resource base and rainfall pattern, they were unlikely to be able to improve their economic prospects.

Another major problem identified in the study, and by the M&E Unit previously, was a generic weakness that lies at the core of many of the other problems associated with the programme. This was the fact that the institutional framework for providing productive services to settlers was fundamentally flawed, or more accurately, simply missing. In particular the arrangements for providing credit, marketing and input supply services were simply inappropriate to meet the needs of settler households. One explanation for this may have been a reluctance on behalf of the original programme planners to follow the Kenyan model of establishing co-operatives to provide these services for fear of setting up organisations which might ultimately involve heavy subsidisation. Alternatively it may have been assumed, mistakenly as it turned out, that the existing service providers, e.g. the AFC (Agricultural Finance Corporation) and the marketing boards, were adequately equipped to cope with expanding service provision to resettlement areas. Whatever the explanation, the lack of efficient service provision eventually led to a vicious circle of increasing indebtedness, low input use, declining productivity and falling economic offtake from settler holdings across many parts of the programme.

These problems undoubtedly contributed to the popular contemporary perception that resettlement was 'failing'. However a number of other factors combined to support this widely held view. These included the valid observation that resettlement was not significantly contributing to resolving the acute problems of overcrowded communal areas. This led government and many donors, including ODA, to re-orientate priority for funding to support development activities in these areas.

But other factors may have been at work which explain why support for resettlement faded away during the late 1980s. For example, it might be argued that after seven years of relatively peaceful post-independence development the 'political' imperative for resettlement had subsided, the issue only coming back on the agenda at election times. Add to this the successful lobbying campaign of the CFU and the political pressure exerted on UK policy makers from individuals and companies with interests in both countries, it is possible to understand why such a resounding endorsement of the economic worth of the programme included in the evaluation report was greeted in many quarters as unwelcome news.

Whilst it would be unfair to suggest that the evaluation report was suppressed by either ODA or the Government of Zimbabwe, it might be argued that neither institution attempted to use it as a basis for re-energising the programme. The general perception of the programme as having been a failure thus persisted.

But the ODA evaluation report was not the only study that perceived the programme positively. The 1991 World Bank *Agriculture Sector Review*, whilst acknowledging weaknesses in the programme accepted that further resettlement would play a central role in the development of the agricultural sector. In 1993 the Comptroller and Auditor General's office undertook a value-for-money study of the programme and concluded that, despite its shortcomings, the programme had benefited some of the poorest people, increased incomes and provided access to services such as clean water and health and education services.

The more positive tone of these and other reports may partially explain the Government of Zimbabwe's increasing impatience over the issue of land acquisition for resettlement during the early 1990s. Whilst much of the motivation was inevitably political in nature, it could be argued that frustration in not being able to unlock some of the economic and social benefits from resettlement, as had been derived during the early 1980s, may have added to this momentum.

However, resettlement was not high on the donor community agenda. There were perhaps two main reasons for this. The first was the widespread and entirely valid recognition that resettlement was having no significant impact on resolving the problems of the overcrowded communal areas. Thus priority for development was quite justifiably deflected towards these areas.

The other reason was perhaps simply the expense of resettlement as practised under the first phase. This was estimated in the report as Z$3,645 per settler household or Z$22,000 in 1996 prices. The original part of the programme financed by the ODA was funded on a 50:50 basis, with the Government of Zimbabwe having to fund the programme up-front before reclaiming half the cost back against the UK grant. This was always problematic for the government, under pressure from other quarters to rein in public expenditure. Whilst programme aid was eventually provided to get round this problem under the ODA part of the programme, the sheer cost of resettlement during the reforming years of the early 1990s rendered it almost impossible for the Government of Zimbabwe to fund it on its own.

This may in part explain why land acquired theoretically for resettlement ended up being leased back to new private tenants as opposed to being resettled by small scale producers during this period. It might be further argued that the disengagement of the donor community from active involvement in resettlement led inevitably to an inability on behalf of the government to sustain resettlement as a well-planned and organised development programme.

Despite these problems it would appear that in the mid-1990s that there was a change of heart amongst the donor community and in particular on behalf of the UK government. The view appears to have evolved, consistent with the findings of the evaluation study, that despite all the operational problems, resettlement can provide economic and social benefits to the poorest members of the population whilst at the same time having an overall positive impact on the economy. Against this background it would appear that the UK and Zimbabwe governments decided jointly to take another hard look at revitalising the programme. This resulted in the Zimbabwe /UK ODA land appraisal mission being mounted in 1996.

The UK Land Appraisal Mission 1996

Despite the UK government's disengagement from direct involvement with resettlement towards the end of the 1980s (in part instigated by the Zimbabwe government's intentions to push ahead with legislation regarding land acquisition which would end the concept of 'willing-buyer, willing-seller' which had been a central tenet of the original programme) ODA agreed to a request by the Government of Zimbabwe to take another look the land resettlement issue in late 1996. This shift in policy on the UK's part was welcomed by Zimbabwe, and in particular within the Ministry of Local Government, Urban and Rural Development, which facilitated the mission.

Prior to the mission team arriving in Zimbabwe, the Ministry had produced a new policy paper on land which outlined government's intentions for carrying out resettlement. This appeared to be a major step forward in that it indicated roles and responsibilities for executing resettlement, indicated different types of land needs for categories of people and described a number of possible ways and means of meeting these needs. Importantly it also made clear that there was a need to retain a substantial commercial farming sector whilst also acknowledging that, regardless of the size of any resettlement programme, the bulk of rural households would still reside in communal areas.

This policy paper was prepared specifically to guide the thinking of the appraisal mission. It was welcomed as a very positive step forward as the paper indicated a pragmatic and flexible approach to resettlement. Specifically the paper identified three categories of land need. The first was for rural landless people, some of the poorest in the country. The second was for skilled and experienced small producers with the capacity to use those skills to develop an important resource productively. The third was for indigenous Zimbabweans wishing to break into the large-scale commercial farm (LSCF) sector. This mix of social, productive and opportunistic objectives appeared to be an ideal starting platform for the work of the mission although the mission was always likely to concentrate on the first two needs as opposed to the third.

One of the most interesting issues stemming from the earlier programme, and carried forward in the new policy document, was that resettlement in Zimbabwe has always been characterised by various 'models of resettlement'. It is not adequately possible to describe all the various models here, but under the new policy it was of interest to note a very important shift in the way future models of resettlement would be planned and managed. The original programme was very much supply-driven. Schemes were planned, established, and settlers selected and moved onto them in a very mechanistic way. Settlers played no role in the design, layout and management of schemes.

Experience elsewhere in Zimbabwe with rural development, particularly in the communal areas, had by the 1990s, moved very much towards a participatory approach to development. This experience and approach had very strongly influenced the new policy paper and had been incorporated into the various models envisaged for future resettlement. Against a background of matching ODA objectives with the Government of Zimbabwe's proposals, the mission identified a potential new phase of resettlement based around three of the models included in the policy paper.

These were the community based Model A programmes, targeted at poor people living in congested areas, for resettlement primarily in natural region III (NR III); *the three tier model*, targeted at providing additional land for livestock based agriculture for low rainfall areas in NRs IV and V; and the *Model A self-contained units* programme, targeted at communal area (CA) farmers wishing to take a step up the farming ladder into small-scale commercial farming, mostly in NRIII and possibly NRII (ODA, 1996). (For further explanation of the Natural Regions refer to Figure 5.1.)

The mission indicated that the participatory nature of the new programme might be further enhanced by linking the planning and management of the programme more closely with the institutional framework covering the provincial and district level planning processes currently being developed and strengthened.

Against the background of a new, broadly identified resettlement programme, the mission examined a range of technical issues which would need to be taken into account under any renewed phase of the programme. A number of these are summarised below.

The issue of land prices, availability and the acquisition process was closely examined: for there to be a programme of any significant size it would be important for land to come onto the market. It was assumed that donors would not provide funds for the compulsory purchase of land although it was recognised that a process approach to resettlement might not fit well with an absolute adherence to the 'willing-buyer, willing-seller' principle. In order to 'encourage' land onto the market without involving a large increase in land prices, measures such as the relaxation of sub-division controls and introduction of a land tax were floated as possible measures. These proposals were guided by the outcome of the Land Tenure Commission, and the Cabinet response to their recommendations.

A second technical issue considered was the institutional capacity to plan and implement a revised programme on any scale. Whereas capacity existed in some functional areas, e.g. physical planning, other areas suffered from severe capacity constraints, e.g. land acquisition. The mission noted that there would be a need to strengthen capacity in certain areas if the proposed new phase was to be implemented smoothly on any significant scale. Much the same observation was made in reference to the provision of productive support services to settler households drawing on the experience of the first phase, in which these services were so badly provided for.

Perhaps unsurprisingly, the mission took another hard look at the economic issues surrounding land resettlement. The mission drew on a number of sources in considering these, including various reports, studies and academic works. On balance, the mission concluded that providing

- confidence could be retained within the smaller remaining LSCF sector;
- that there was underutilisation of land within the existing LSCF sector;
- that there would be intensification of land use under resettlement using less capital intensive production

then there would be significant economic gains to be made from resettlement. However, given the lack of hard data on the issue, the mission indicated that these assumptions might need to be validated through further research.

Other technical issues addressed during the mission included social development and gender issues, consideration of the position of commercial farm workers under resettlement, tenure issues and environmental issues. All of these need to be fully considered at the design stages of a revised programme.

A further key issue considered during the mission was that of the cost of any new phase of resettlement and how and by whom this cost might be met. Some indicative costings were made by the mission and included in the report. It was estimated that a programme covering five years and 25,000 households would cost Z$1.5 billion rising to Z$2.3 billion for a programme covering 35,000 households. The cost estimate per settler household would be over Z$60,000. It was noted that this was much higher than the first phase estimate which worked out at Z$22,000 at 1996 prices. The main factor pushing up the price appeared to be the price of land, with an additional factor being the level of development expenditure on prescribed levels of infrastructure. Whilst it was recognised that this was a very expensive and ambitious programme it was considered that much of the development costs might well be funded from external sources. The mission consulted most of the major donors who indicated interest in being involved with a forum to consider how the revised programme might be funded. The mission understood at the time that the UK, and possibly one or two other donors might even assist with land purchase. Others were positive enough to give the impression that funding development costs could be a possibility. However, the cost of administering and running the schemes would all have to fall on the domestic budget which, given the scale of the programme envisaged, may well have been a limiting factor.

In summary, it is fair to conclude that, having re-visited many of the key issues, the mission came up with a report which was very positive in tone and which clearly indicated a way forward. The report included recommendations for undertaking a number of further studies to run along side an inception phase for the revised programme, during which the programme was to be designed in greater detail and during which funding sources other than ODA would be identified.

As it turned out, it would appear that this approach may have been considered too tentative by the Government of Zimbabwe. Expectations may have been raised beyond what might reasonably have been achieved through the execution of such a short mission. Furthermore, whilst the mission report was considered by the mission team to have been closely in line with the policy of the Government of Zimbabwe, it may be that it could have been interpreted differently. Whether or not this was the case, the government's response to the mission report was slow to emerge, by which time further pressure on the land issue had built up, resulting more recently in a more worrying and direct approach to land redistribution.

The main conclusion that might be drawn from the output of the two studies discussed above is that in order for the economic benefits of resettlement to be maximised and sustained, it will be essential for the international community to provide finance and other support to the Government of Zimbabwe, so that it can manage the process of land redistribution in an efficient and effective manner. The alternative is a confidence sapping 'land-grab', from which few can benefit.

References

Adams, M., Cassidy, E., Cusworth, J., Lowcock, M. and Tempest, F. (1996), *ODA Land Appraisal Mission Report*, British Development Division in Southern Africa, Harare.

Comptroller and Auditor General, Government of Zimbabwe (1993), *Value for Money Project (Special Report) on Land Acquisition and the Resettlement Programme*, Harare.

Cusworth, J. and Walker J. (1988), *Evaluation Report EV434*, Overseas Development Administration, September 1988.

Cusworth, J. (1991), Land Resettlement Issues, Background Paper for *Agricultural Sector Review*, World Bank, Harare.

Land Tenure Commission (1994), *Report of the Land Tenure Commission*, Harare.

4 The Politics of Land Reform in Zimbabwe

PROFESSOR LIONEL CLIFFE

The period immediately after independence in 1980 saw a major policy initiative for agrarian reform. Under the Land Reform Programme (LRP) three million hectares of large commercial farms (LSCF), previously owned and farmed by whites, were eventually transferred to African smallholders. Now, twenty years on, a second land reform was about to be implemented (Government of Zimbabwe, 1999), even before pre-election land invasions in 2000. This contribution explores the political forces that were at work in shaping past policies and the context for their implementation, generating constraints that among other things explain why and how the ambitious targets set at the outset were not realised.

This chapter concentrates on the experience of the first three to four years which were marked by a rapid process of land acquisition by the government for resettlement (virtually all the land used for resettlement up to the mid-1990s was in fact acquired in 1980-82), and by a rapid acceleration in the numbers resettled (from a slow start in 1981 almost half of all those resettled in the 18 years since independence were resettled in 1982-84), and the building up of institutional capacity to handle such a programme. But this period was also characterised by a second defining element: criticism of the programme associated with severe cuts in the budgetary provision for it, and eventually in the cutting back of the institutional capacity for implementing new resettlement schemes. Some of the political circumstances and consequences curtailing the LRP are worth examining with the hindsight that is offered from a whole range of Zimbabwe government, British ODA (Overseas Development Administration) and scholarly assessments (Government of Zimbabwe, 1987; ODA, 1988 and 1996; Moyo, 1995; Kinsey, 1999) and several contributions to this volume, which show that the severe criticisms of the early 1980s were misinformed, premature and not borne out by later evidence. More concretely, this experience raises the question as to whether and how such political undermining of a future programme can be avoided.

These episodes will be primarily explored in terms of an interplay of various and often conflicting stances among those in a position, inside and outside Zimbabwe, to influence policy. Such inter-elite political struggles had, in fact, shaped and reshaped land policy at several key moments in the history of the former Rhodesia and after. But they have been played out against broader social and political forces.

Colonial Background

To situate these broader forces and elite interests, brief reference must be made to the inherited structures of land relationships in Rhodesia and Zimbabwe. The 'dualist' structures of agriculture, whereby in one sector one per cent of the farmers owned half of the agricultural land and by and large the more productive half, while 700,000 farming households (98 per cent) had the other half in what are now called the communal areas (CAs), is familiar. However, the structure of each, and their interlinkages, were systematically shaped and reshaped by state intervention throughout this century by successive colonial, settler-colonial and post-colonial regimes. Examples of the interconnectedness of the two sectors and their relative prospects include: wholesale forced removal of African people to make room for white farming, which continued up to the 1950s; ensuring the availability of cheap farm labour for the large farms that was 'reproduced' in the CAs; the elimination and prohibition of any competition from African smallholders producing on any significant scale for the market; land tenure systems in the CAs which maximised (at times) the absorptive capacity of these 'reserves' and ensured social control mainly through chiefly power to allocate land. Brief note needs to be made of the period in the 1950s when new 'liberal' legislation promoted individual titling of land in the CAs, along with a set of conservation measures, as these policies generated wide-scale rural protest from both those dispossessed or those whose rights to land were made more vulnerable, and the chiefs who lost specific powers and general authority thereby. With the coming to power of the white-supremacist government of the Rhodesian Front, 'traditional' land tenure systems were reimposed, at the same time as the Unilateral Declaration of Independence (UDI). The UDI regime thereby called a halt to efforts to incorporate, politically and economically, an urban middle class and a rural small commercial farmer class of Africans, preferring a strategy where its only black allies were the chiefs, re-empowered to allocate land.

One further dimension of this longer historical trajectory was that the several ways in which politically dominant interests resolved the 'land issue' at any one moment was often a basic focus for political protest and resistance. Thus the resurrected 'native' administrations under the UDI regime, even with the tool of control of land allocation, were not able to stem the tide of discontent and growing political consciousness, hence tensions rose throughout the 1960s. These translated into demands about land, which in turn were further focused, as guerrilla methods of struggle by the liberation movement spread from the eastern borders with Mozambique from 1972. As many analysts of the *Chimurenga* liberation struggle have pointed out (Ranger, 1986), the call for land redistribution was a major factor in mobilising popular support. The result of the war was not merely that expectations for land reform were high. At the time of independence in 1980, there had already been an actual process of removing white occupants from farms, especially on the fringes of the white-owned heartland. Some embattled farmers were chased off the land, or retired temporarily to town, or to South Africa. Fences of ranches were cut for peasants to graze by 'poaching'. Much of the land that was formally 'acquired' and then 'resettled' in the early 1980s was often regularising such *de facto* occupations that had occurred during or just after the war, by structuring land use and providing services on these areas. One crucial origin of the dynamic for land reform was not just the broad forces of popular opinion but such specific instances of people voting with their feet. A whole range of popular initiatives can be seen as both a form of social and political pressure but also, as Moyo (1995) has stressed, as symptoms of *needs* for land - although sometimes by land entrepreneurs and not always the landless or land poor. People 'squatted' on farms, on unutilised land, cut down forest, occupied areas, like *dhombos* (wetland depressions) or river banks, which were 'protected' for conservation. Groups cut wire or grazed by dead-of-night on commercial ranches or other range, made 'deals' to borrow, lease, buy or have assigned to municipalities land that would provide elbow room.

Early Resettlement Performance

As indicated in the introductory paragraph, the period 1981-83 saw a dramatic expansion in resettlement activity. At the time of independence in 1980 the Government of Zimbabwe announced a target of settling 18,000 households on 1.1 million hectares of land. By 1983 it had acquired far more than that target area - 2.2 million hectares. Most of the total of 3.8 million hectares that was acquired by the 1990s was obtained in those early years. From small beginnings in Fiscal Year 1980/81, almost 10,000 and 15,000 households were settled in FY1981/82 and 1982/83 respectively. This latter figure was never to be even remotely approached in any year thereafter. It fell back to c.2,500 in 1983/84, recovered significantly to 8,000 plus in 1984/85 but never surpassed 2,500 to 3,000 in all subsequent years. Meanwhile, in announcing its Transitional Development Plan, 1983-85, the government opted for a much expanded goal of settling 162,000 households - presumably as the eventual target, not for that 3-Year Plan. It further specified an annual target of 15,000 for the First Five-Year National Development Plan, 1986-90. But the slowdown had already begun, so while almost 40,000 households had been resettled by 1985, the total reached by 1990 was only 48,000, and 71,000 by 1997.

In explaining this shortfall between performance and targets, it is worth recognising that the rate of 15,000 was in fact realised in one year during this peak period. So any arguments that the targets were inherently 'unrealistic', that there was not the capacity to deal with that number of households, have to be more closely examined. Indeed such explanations do not uncover reasons behind the sharp fall-off in new resettlement in 1983/84 nor the very much lower rate after 1985/86.

There are, in fact, several explanations on offer as to why ambitious targets were not realised. In reviewing them it must be realised that for any explanation to be satisfactory, it must also address reasons for this decline in progress after 1984. One often cited assertion is the 'constitutional constraint which has had a debilitating effect on policy formulation' (Mumbengegwi, 1986). Specifically, the 'Freedom of Deprivation of Property' provisions of the Bill of Rights, agreed at the Lancaster House Conference in London in 1989, whereat British chairing 'provided imperialism with the opportunity to be an 'umpire' in a 'match' in which it had a vested interest', specified land could only be compulsorily acquired if it was underutilised and compensation was 'adequate' (as defined by the market) and 'prompt' - and in foreign exchange. This effectively reduced the independent government's options to acquiring land on a voluntary basis, when it could pay for land offered it in Zimbabwe currency, and/or relying for funding on foreign donors. Some commentators suggest that this central constraint was buttressed by broader elements of the compromise, leading to independence which meant the new state was intent on keeping Western powers, international financial institutions and neighbouring South Africa and, in turn, the white settlers, 'sweet'. But unlike those who simply specify these inherited circumstances as representing an insurmountable obstacle to any radical land redistribution from the outset, Mumbengegwi (1986) does provide an argument as to how the restrictions only began to bite later:

> initially, abandoned (as a result of the liberation war) and underutilised land was the prime acquisition target...(but) after the first three years, 'willing-buyer, willing-seller' land tended to dry up and prices rose significantly, due to a combination of better agro-ecological conditions and unavailability of willing sellers.

While it is certainly true that much of the sizable area of land acquired between 1980 and 1982 was 'available' as a result of the actual patterns of fighting on the ground, the reliance on the turning off of the supply of land as the sole explanation of curtailing resettlement does not fit some of the evidence. There remained, for instance, several hundred thousand hectares of 'acquired' land that remained unresettled until the late 1980s. Even allowing for the fact that much of this was in areas of the Matabeleland South province where there had been fighting and disturbances, and for the fact that some acquired land was what some Resettlement Officers at the time disparaging referred to as 'baboon country', there was some margin for continued resettlement.

Another specific constraint cited was the limited availability or drying up of donor funds. That this was the decisive restraint is gainsaid by the fact that some earmarked British funds remained untouched until after 1990. The formula by which the UK government released funds was that it provided half of projected expenditure for each resettlement scheme, both for the land acquired and for development costs, once plans had been detailed and approved. Thus donor funds were drawn down automatically to match Government of Zimbabwe funding of projects. It was those latter allocations which were reduced, from a relatively early stage. The exchequer vote for the Ministry of Lands, Resettlement and Rural Development was reduced slightly for the financial year 1983/84, and more substantially thereafter. There were clearly budgetary constraints. Drought in 1982 compounded emerging balance of payments problems as the end of sanctions allowed for replacement of equipment and investment. But it was political pressures that influenced the government to curb the resettlement budget provision disproportionately.

Internal Opposition to Resettlement

One avenue through which sentiments critical of the programme were articulated was the Parliament's Estimates Committee. Under the chairmanship of Mr. Cartwright, a Rhodesian Front MP and a farmer, it reviewed the budget of the Lands & Resettlement ministry and produced a report highly critical of the programme. Many of the criticisms of the committee were trivial: houses on schemes were 'dirty'; Resettlement Officers were not there to meet them - but their antipathy gave apparent sustenance to a stereotyped view that resettlement schemes were failures, which was widespread among some opposition and government politicians. However, in addition to those familiar elements of that ideology, such as the threat of environmental degradation (see Cliffe, 1988) and the unproductive character of African small-holders as givens, the Estimates Committee came up with another assertion which was to prove effective in undermining the resettlement effort. They characterised the Resettlement Officers, who were key catalysts and coordinators on the ground, as 'out of control' and as a 'law unto themselves'. These vague accusations which were pursued behind the scenes eventually pointed a finger at the resettlement programme's dynamic director, Sam Geza, who was charged with exceeding his authorised powers and misuse of funds, and was suspended. After several months of investigation, during which the Directorate's energies were diverted from implementing resettlement, he was completely exonerated by the Public Service Commission, and indeed congratulated for his work and promoted eventually to permanent secretary status. But the damage had been done: the Directorate had been discredited, its efforts had been diverted and its budgets cut. The implementational capacity was thereby reduced and was never regained. Indeed, one of the issues faced now in Phase 2 of an accelerated land reform programme is the need to quickly re-establish such a capacity. But perhaps even more serious was the consequence for the perception of land reform: it tended to be seen in official circles as something of a failure, a temporary concession on which the book could now be closed. This attitude spread to public opinion and to audiences outside. It was even to influence approaches to land reform in South Africa. Zimbabwe experience was explicitly referred to as a failure and irrelevant to South Africa, making it easier for World Bank-influenced approaches based on 'market-based' reform to prevail, with adverse results on land reform there.

Jenkins (1997) identifies such 'CFU... lobbying hard against resettlement, through Parliament, for instance, as early as 1980. But what needs to be further explained if this account of the slowing down of resettlement is to be complete is why other elements in government did not effectively resist these pressures, especially in a context of widespread popular pressure and expectations about land.

One common explanation is to suggest that the shared interests of existing white farmers and aspirant African land owners was characteristic of what even a mainstream Western liberal commentator sees as a broader 'alliance between dominant representatives of the state and black capitalists' (Kinsey, 1999), replacing 'the old (pre-independence) compact between government and the white captains of industry and agriculture'. In fact, as we have seen, even the old compact was liable to shifts: with UDI a period of the dominance of the interests of 'white captains of industry' and metropolitan and broader international capital gave way to a more old-fashioned political predominance of white agriculture and other settler interests. Other analysts see the 'post-settler capitalist state', in Mandaza's phrase (1986), as reflecting these international and local capitalist interests more directly and not just those of the emerging African bourgeoisie and petty bourgeoisie. Herbst (1989) offers a more nuanced explanation of the outcome of agrarian policy and practice in the 1980s which recognises the coalitions of interests without seeing them as over-determinant, thus providing a place for the ideas and perceptions of policy-makers. He argues that what eventually inhibited any such inclination was that government did not develop a 'long-term vision of how agriculture in Zimbabwe will develop', and the absence of the necessary self-confidence of an alternative vision which could have challenged the existing orthodox of growth dependent on keeping large commercial farms intact. These factors reinforced a 'risk-aversion strategy', with the result that 'organizational and technocratic decisions are allowed to dominate Zimbabwe's land policy'. A further factor in his explanation of that outcome is that despite the general popular sentiments, 'peasant farmers are unable to bring significant political pressure on the government' (Herbst, 1989).

One piece of evidence that Herbst offers for this last factor is the inability of those voices to counter an argument put forward increasingly in the late 1980s by the representatives of African small-scale commercial farmers, and supported by the CFU. This was that 'master farmers' and agriculture institute graduates and others like themselves, rather than peasant farmers or the landless should be the ones to be resettled. Indeed, this had become a position broadly accepted within government, and was reflected in LRRP-2 proposals that a significant proportion of resettled land should be for such beneficiaries. This position remained despite evidence from the ODA Report (Cusworth & Walker, 1990) that differential 'success' among those resettled in the 1980s was a reflection of their initial assets rather than 'ability', and the even-stronger, statistically 'robust' finding (Gunning *et al.*, 1999) that 'differences in initial conditions across...households, such as previous farming experience, have few persistent effects'. The counter-argument for maintaining the original aims of providing opportunities for the poor, rather than small or large-scale African commercial farmers, is given weight by findings that 80 per cent of the 1,200 indigenous commercial farmers were facing serious indebtedness and their representatives were suggesting that a solution might lie in some of their holdings being bought by government for more conventional resettlement (*Sunday Mail*, 13 June 1999). This was an implicit acknowledgment that, like white farmers, many could not fully utilise the sizable hectarages that comprise the average farm.

International Involvement in Land Reform

International forces have consistently helped to shape land reform policy, whether through the broad, anonymous influences that define whether global capitalism has 'confidence' in the Zimbabwe economy, through specific diplomatic initiatives by global or regional powers, or through the rural programmes of aid agencies. Henry Kissinger, on behalf of the US government, came up with a proposed British-US 'Zimbabwe Development Fund' of the order of US$1-1.5 billion (UK, 1977) to be used 'to facilitate the economic transition while minimising its disruptive effects on ... growth'. Its proposed share was never to be remotely met by any US assistance to independent Zimbabwe, although the UK did provide funding for resettlement, as we shall see. Of course the proposals that came out of Geneva were rejected by the Rhodesian Front government and the war continued for another three years.

By the time of the 1979 Lancaster House Conference, the idea of a Fund had disappeared. In those negotiations the land issue was perhaps the most difficult one to resolve. The Patriotic Front delegation, the coalition of ZANU and ZAPU, rejected the UK proposals on land well into the negotiations in October, thereby making a stand on what was by then the only outstanding issue. The talks broke down and the UK Foreign Secretary, Lord Carrington, suspended the Patriotic Front from Lancaster House. Behind the scenes, pressure from the west on the front-line states, and by them on the Patriotic Front, combined with the US 'coming to the rescue' (Tamarkin, 1990) by making a commitment to what Stedman (1991) termed 'ambiguous support' for a fund that would help Zimbabwe bear the economic burden of compensation for land, as well as pensions for Rhodesian officials.

Certainly as far as the US was concerned the support was 'ambiguous' as no contribution for such purposes was made when Zimbabwe became independent the next year. The UK did however make an initial commitment to a resettlement programme by a grant of £20 million in 1980 and a further £10 million in 1981, with a further £17 million of programme aid loans being agreed in the 1983-85 period. The grant agreements specified that release of such funds for specific resettlement projects should be matched by a similar allocation by the Government of Zimbabwe, on a 50-50 basis. In the event projects were completed at a lower cost in terms of pound sterling than planned, mainly as a result of the devaluation of the Zimbabwe dollar. As already noted, the Government of Zimbabwe's budgetary cutbacks to the programme meant that the UK committed funds were not drawn down; indeed, it was estimated that £3,382 million of the original 1981 grant remained unspent in September 1996 (ODA, 1996).

Land Reform and Resettlement - Phase 2

In 1998 a process of formulating a second stage of land reform began. It was more broadly-based across government than the earlier policy and also involved consultation with a wide range of 'stakeholders'. The plan (Government of Zimbabwe, 1999) envisaged that a further 5 million hectares of land would be acquired from the LSCF, (i.e. 50 per cent more than in the 1980-97 period, to resettle about 150,000 households, envisaging a more intensive allocation of land than before). Much of this land will be designated for compulsory acquisition with compensation, but will also include voluntarily offered land, including subdivisions of existing holdings - a flexible approach often urged in the past (Marongwe, 1999; Cliffe, 1986) to take advantage of the widespread phenomenon of underutilised land on generally productive farms. This LRRP-2 will initially build on the approach used in the past, i.e. for the most part, acquisition of land by government which will in turn subdivide farms into small-holdings, along the lines of the Model A resettlement schemes. But it will provide settled households with more secure title to land allocations than the 'permits' previously granted, and will provide for a more participatory institutional structure, which are welcome applications of lessons learned.

This substantial, imaginative and practicable plan seems to have secured the support of a broad spectrum of stakeholders. A broad range of governmental departments have been involved in the planning and have involved extensive consultations with specialists and academics, Zimbabwean NGOs and farmers' groups. The views of such interested groups and government and private analysts suggests that a self-assured and coherent alternative challenge to 'white settler agricultural orthodoxies' has emerged among some Zimbabweans - unlike the situation in the 1980s. The still largely white Commercial Farmers Union seems to have given broad support to the proposals, even if somewhat dubious about their extent. In fact they were always in favour, for a mixture of reasons of political self-preservation and economic self-interest, of a limited redistribution provided they could shape it. But this distinct change of emphasis probably stems from several factors, including the fact that the white farmer class has not reproduced itself demographically, as sons have gone into other occupations or countries, and the tendency for the most commercially dynamic farmers to switch to high-price, intensive cultivation of new horticultural export products like flowers (now the second largest agricultural export) and fruit and vegetables, whose land requirements are small. There may therefore be less of a tendency for such vested interests to undermine future land reform than in the 1980s, although they will still seek to affirm the supremacy of the large farm sector, even a diminished and more intensive one, and existing and would-be black farmers' interests may be even less easily accommodated within land reform to benefit the poor. Several other overseas donors as well as the British have indicated a willingness to support this LRRP-2 - despite the shrill, and often counterproductive sloganising of President Mugabe. But if the obstacles from internal classes and foreign capital have perhaps lessened, will this translate into sufficient financial support for a land transfer still meeting their requirements for compensation? The major question remains: whether those in the highest levels of government have the long-term resolve, even under continued pressure from the landless and small peasants which has not abated, and even after the land invasions that they have orchestrated for short-term electioneering reasons, to commit the resources and build up and maintain the institutional capacity to implement such a plan fully.

References

Cliffe, L. (1986), *Policy Options for Agrarian Reform in Zimbabwe: A Technical Appraisal*, Paper submitted by UN Food & Agriculture Organisation to Government of Zimbabwe.

Cliffe, L. (1988), 'The Conservation Issue in Zimbabwe', *Review of African Political Economy*, 42, 40-48.

Cusworth., J. & Walker, J. (1988), *Land Resettlement in Zimbabwe: A Preliminary Evaluation*, Overseas Development Administration, London.

Government of Zimbabwe (1987), *Report on the National Symposium on Agrarian Reform in Zimbabwe*, Dept. of Rural Development, Ministry of Local Government & Rural Development, Harare, December.

Government of Zimbabwe (1998), *National Land Policy: Framework Paper*, Ministry of Lands & Agriculture, Harare.

Government of Zimbabwe (1999), *Land Reform & Resettlement Programme, Phase 2: Inception Phase Framework Plan, 1999-2000*, Technical Committee of the Inter-Ministerial Committee on Resettlement & Rural Development and National Economic Consultative Forum Land Reform Task Force, Harare.

Gunning, J., Hoddinott, J., Kinsey, B. and Owens, T. (1999), *Revisiting Forever Gained: Income Dynamics in the Resettlement Areas of Zimbabwe, 1983-1997*, Working Paper WPS/99-14, Centre for the Study of African Economies, Oxford.

Herbst, J. (1989), *State Politics in Zimbabwe*, University of Zimbabwe Press, Harare.

Jenkins, C. (1997), 'The Politics of Economic Policy-Making in Zimbabwe', *Journal of Modern African Studies*, 35, 4.

Kinsey, B. (1983a), 'Emerging Policy Issues in Zimbabwe's Land Settlement Programmes', *Development Policy Review*, Vol. 1.

Kinsey, B. (1983b), 'Forever Gained: Resettlement & Land Policy in the Context of National Development in Zimbabwe', in J. Peel & T. Ranger (eds), *Past & Present in Zimbabwe*, Manchester University Press, Manchester.

Kinsey, B. (1999), 'Land Reform, Growth & Equity: Emerging Evidence from Zimbabwe's Resettlement Programme', *Journal of Southern Africa Studies*, 25, 2.

Mandaza, I. (ed) (1986), *Zimbabwe: The Political Economy of Transition, 1980-1986*, CODESRIA, Dakar.

Marongwe, N. (1999), *Civil Society's Perspective on Land Reforms in Zimbabwe: Some Key Suggestions from a Survey*, ZERO Regional Environment Organisation, Harare.

Ministry of Lands, Agriculture & Rural Settlement, 1986, *First Annual Survey of Settler Households in Normal Intensive Model 'A' Resettlement Schemes, Main Report*, Monitoring & Evaluation Section, Central Planning Unit, Harare, September.

Moyo, S. (1986), 'The Land Question' in I. Mandaza (ed.) *Zimbabwe: The Political Economy of Transition, 1980-1986*, CODESRIA, Dakar.

Moyo, S. (1995), *The Land Question in Zimbabwe*, SAPES Books, Harare.

Mumbengegwi, C. (1986), 'Continuity & Change in Agricultural Policy', in I. Mandaza (ed.) *Zimbabwe:The Political Economy of Transition, 1980-1986*, CODESRIA,Dakar.

Overseas Development Administration (1996), *Report of ODA Land Appraisal Mission to Zimbabwe*, British Development Division in Central Africa, ODA, London.

Parliament of Zimbabwe (1983), *Third Report of the Estimates Committee on the Ministry of Lands, Resettlement & Rural Development Vote No. 20 - 1982-83*, House of Assembly, Government Printer, Harare.

Ranger, T. (1985), *Peasant Consciousness & Guerrilla War in Zimbabwe*, James Currey, London.

Ranger, T. (1993), 'The Communal Areas of Zimbabwe', in T. Bassett & D. Crummey, eds. *Land in African Agrarian Systems*, University of Wisconsin Press, Madison.

Stedman, S. (1991), *Peacemaking in Civil War: International Mediation in Zimbabwe, 1974-1980*, L. Rienner, Boulder.

Sunday Mail (1999), 13 June 1999.

Tamarkin, M. (1990), *The Making of Zimbabwe: Decolonisation in Regional & International Politics*, F. Cass, London.

UK Government (1977), *Rhodesia: Proposals for a Settlement*, Presented to Parliament by the Secretary of State for Foreign & Commonwealth Affairs, September, Cmnd. 6919, HMSO, London.

Weiner, D. (1989), 'Agricultural Restructuring in Zimbabwe & South Africa', *Development & Change*, 20, 3.

5 Zimbabwe Land Policy and the Land Reform Programme

EDITED BY DR COLIN STONEMAN

This paper summarises the Government of Zimbabwe's perspective on land reform issues, as presented at the 1998 SOAS conference. Its contents are derived from a press statement issued by the Zimbabwe High Commission in London in December 1997 and circulated at the conference, and from the presentations and contributions made by representatives of the Zimbabwe government at the 1998 conference. Information derived from the press statement is *verbatim*, and clearly referenced in the text. Other information derives from the transcript of the conference presentation, and is also clearly referenced as such in the text. As with any transcript of a conference, which includes unstructured discussion such as questions and answers, the transcript itself does not lend itself to published form. Every effort has been made to fully and neutrally represent the main messages of the material presented. While responsibility for compiling this material ultimately lies with the editor of this chapter, all such material is transparently reflected in the press statement, the conference transcript, and the overheads made available to the editor by the representatives of the Government of Zimbabwe. It should be pointed out that the emphasis of conference discussions was often audience-led, hence information does not necessarily reflect the representative's own emphasis on the range of issues related to land reform.

The Zimbabwe Land Issue in Perspective

The need to provide land to the landless has been central to the Government of Zimbabwe's efforts to raise the standard of living for the deprived majority in over-crowded, over-stocked and over-used communal areas. Land has always been a central issue in Zimbabwe. The liberation struggle that ended with the Lancaster House Conference Agreement was essentially about land and during that Conference, land acquisition and distribution were major areas of contention, at one point almost wrecking the negotiations. The final agreement failed to adequately address the problem of land and its distribution to the landless peasants. For that reason, the Zimbabwe government enacted the Land Acquisition Act of 1991 which was passed by Parliament and is the basis on which land acquisition and resettlement is managed. The 'willing-seller, willing-buyer' principle enshrined in the Lancaster House constitution which was applicable during the first decade of Independence, has since been superseded by the above mentioned Act (Zimbabwe High Commission, 1997).

Key elements and institutions of the National Land Policy (NLP) are summarised below.

Key elements of National Land Policy (NLP), as stated in the Land Acquisition Act of 1991

- Preamble and historical background
- Programme objectives and strategies
- Legal framework
- Administrative arrangements
- Institutional framework
- National Land Board
- National Land Trust
- Settlement Models
- Land Valuation and Compensation
- Procedures
- Tenure systems
- Land tax
- Indiginisation
- Redistribution procedures

Source: Takavarasha, 1998

Agroclimatic Zones of Zimbabwe

The agricultural potential of Zimbabwe's lands is classified according to the scheme in Figure 5.1. The map shows the localities of these zones across Zimbabwe (adapted from Surveyor-General, 1984, *Zimbabwe: Natural Regions and Farming Areas.* 1:1,000,000).

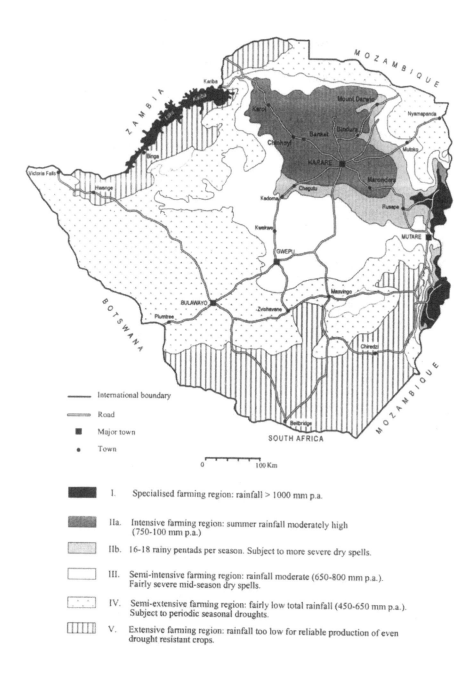

International boundary

Road

Major town

Town

0   100 Km

I.   Specialised farming region: rainfall > 1000 mm p.a.

IIa. Intensive farming region: summer rainfall moderately high
      (750-100 mm p.a.)

IIb. 16-18 rainy pentads per season. Subject to more severe dry spells.

III. Semi-intensive farming region: rainfall moderate (650-800 mm p.a.).
      Fairly severe mid-season dry spells.

IV.   Semi-extensive farming region: fairly low total rainfall (450-650 mm p.a.).
      Subject to periodic seasonal droughts.

V.   Extensive farming region: rainfall too low for reliable production of even
      drought resistant crops.

Figure 5.1 Zimbabwe: Natural Regions and Farming Areas

Distribution of Land in Zimbabwe

Figure 5.2 shows the distribution of land at a) Independence (1980), b) at present (1998) and c) projected, according to government plans. Table 5.1 shows the same data in tabular form.

a) Distribution of land in Zimbabwe 1980

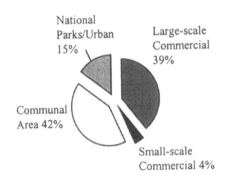

b) Present distribution of land in Zimbabwe

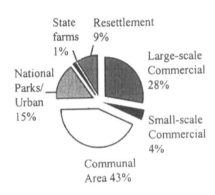

c) Proposed land distribution pattern

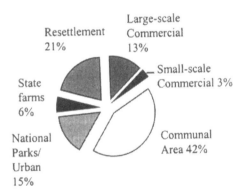

Figure 5.2 Land distribution patterns in Zimbabwe

Source: Takavarasha, 1998

Table 5.1 Distribution of land in Zimbabwe, past (1980), present (1998) and projected

	1980	*1997*	*The future*	
	%	*%*	*%*	*mn ha*
Large-scale commercial	39	28	15	6.0
Communal areas	42	43	42	16.4
Small-scale commercial	4	3	3	1.2
National parks/urban	15	15	15	6.0
Resettlement areas	-	9	21	8.3
State farms	-	1	4	1.4

Source: Takavarasha, 1998

The acquisition and resettlement programme which followed Independence, based on the 'willing-seller, willing-buyer' principle saw the transfer of 9 per cent of LSCF to resettlement schemes. The purchase of these lands was in part aided by the British government, which has made available 20 million pounds since 1980, in support of the resettlement programme (Zimbabwe High Commission, 1997). Most of this land - 3.3 million hectares - has been resettled. Land which has been acquired but not resettled is held by the Agricultural and Rural Development Authority and leased out for grazing and cultivation purposes. Hence, all the land that has been acquired is being utilised, the majority under resettlement schemes and the rest under leasehold (Conference transcript, 1998). Despite the initial resettlement efforts, the current land distribution in Zimbabwe (a country of almost 12 million people) remains skewed in favour of large commercial farmers, as demonstrated by the following indices (Takavarasha, 1998).

- 4,000 large scale commercial farmers (mainly white) on 11.2 million hectares;
- More than 1 million communal area families on 16.3 million hectares (mainly in drier and less fertile agroclimatic regions);
- 10,000 small scale commercial farmers on 1.2 million hectares;
- 70,000 resettlement families on 2 million hectares;
- State farming sector on 0.5 million hectares.

Objectives of the Land Reform Programme

The following objectives are emphasised.

Equity

The Land Acquisition Act of 1991 (Government of Zimbabwe, 1991) was enacted by the Government of Zimbabwe to address the prevailing inequities of land distribution in Zimbabwe. The government carried out a land distribution exercise which determined that 5 million hectares were needed to resettle the landless. The land required for resettlement purposes was identified on the basis of criteria outlined below:

1. land from derelict and/or under-utilised properties;
2. land owned by absentee landlords;
3. land from farmers with more than one farm or with oversized farms;
4. land contiguous to communal areas.

On 28 November, 1997, the government published a gazette listing some 1,480 farms which it intends to acquire in terms of the Land Acquisition Act of 1991. It has been decided that land should be distributed more equitably. No farmer will be without land in Zimbabwe (Zimbabwe High Commission, 1997).

Efficiency

Recent media reports have wrongly suggested that the envisaged land reform programme will destroy agriculture in Zimbabwe. They overlook the fact that the new land tenure system will actually encourage a more intensive utilisation of arable land. Besides, the reports have not reckoned with the well-known fact that the small-scale farmer in Zimbabwe has, through government's resettlement input support policies, moved from virtual agricultural subsistence at Independence to the current position where his/her maize, cotton and groundnuts output accounts for over 70 per cent of national output. The communal farmer has been able to achieve this under severe land limitations, the average family holding in that sector being a mere 5 acres which gets progressively subdivided with the growth of the family. Given that global hectarages under these crops has declined from the mid-1980s, as a result of large commercial farmer diversification into activities such as game ranching and tour and safari operations, it follows that the small holder sector operates at better and higher levels of productivity than the former and today underpins the food security in the country and region as a whole (Zimbabwe High Commission, 1997).

Economic Growth

The Government of Zimbabwe is committed to agricultural development. The majority of Zimbabweans derive their livelihood from agriculture and live in the rural areas where the need for land is greatest. Providing land to the landless is the most effective way of eradicating poverty. Thus land acquisition must not be regarded as an end in itself but must be seen as a means of improving the efficiency of agriculture and raising the standard of living of all Zimbabweans (Zimbabwe High Commission, 1997).

Peace and Stability

For Zimbabwe, equitable land distribution is vital not only for economic development, but also for national peace and stability. Just as economic development is fragile in an unstable political environment, so too is social stability dependent on equitable economic institutions. The objectives of long-term peace and stability in Zimbabwe are directly served by a more equitable distribution of land in the country (Conference transcript, 1998).

Components of the Land Reform Programme

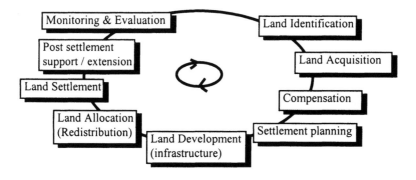

Figure 5.3 Components of the Land Reform Programme

Source: Takavarasha, 1998

There are nine components of the Land Reform Programme, illustrated in Figure 5.3, above. The first stage of the process, involving the identification of land for resettlement (5 million ha), has now been undertaken on the basis of the criteria noted in the previous section (p.52).

Institutional Roles

Implementation of the remaining components of the Land Reform Programme will be based on a 'holistic' approach by which all stakeholders in land-related matters can contribute to the process. These contributions need not supplant the government's programme but rather, will constitute complementary inputs into better ways of continuing the programme (Conference transcript, 1998). The implementation of the remaining components require the support of various institutions, including:

Government departments:
       Interministerial Committee on Resettlement and Rural Development
       Department of Agricultural Technical and Extension Services
       (Agritex)
       Agricultural and Rural Development Authority (ARDA)
       Agricultural Finance Corporation (AFC)
Non-government bodies:
       Non-governmental agencies
       Farmers Unions
       Banking Sector
       Agri-Businesses
       Legal firms
       Agricultural consultancy firms
       International Donor Agencies

Source: Takavarasha, 1998

Key Issues

Farmer Support Services

Key to the economic success of resettlement is the ability of farmers to meet productivity targets within the shortest possible time-frame (Conference transcript, 1998). The implementation process therefore requires the coordination of institutions in supporting resettled farmers in terms of the following provisions:

- Land use planning
- Administration of lease agreements and permits
- Farmer training
- Agricultural extension
- Credit and input supply
- Product marketing
- Irrigation infrastructure
- Farm budgeting
- Performance analysis

Source: Takavarasha, 1998

Funding

In order to support resettlement, the required resources must be planned alongside the implementation process. The resulting costs may well be beyond the capacity of the British government to assist, and beyond the capacity of the Government of Zimbabwe to finance. However, consensus between the British government and the Zimbabwean government would greatly encourage the donor community to lend assistance to an appropriate and well-designed reform programme (Conference transcript, 1998).

Ownership

Land ownership should be for Zimbabwean nationals. Where foreign investment programmes require land purchase, consideration will be given, though this situation has not yet arisen. Regarding ownership by resettled farmers, there is the Commercial Farm Resettlement Scheme, as well as the current leasing of land under government and parastatals. Regarding women's ownership rights, a law on inheritance was recently passed which replaces the eldest male as heir, making the heir the surviving spouse and children. Through our Age of Majority Act, and through the education of women by NGO's, regarding their entitlements, we are moving towards more equitable entitlements for women (Conference transcript, 1998).

Communal Areas

The issue of communal area (CA) re-organisation is a major task. It will involve both the ministries of Land and Agriculture and Local Government. Re-organising and re-planning the structure of the villages, the infrastructure that supports the villages and the support needed for production purposes, is an ongoing programme in Zimbabwe. Communal areas are not intended to be refugee areas for the poor. Once the CAs have been decongested and re-organised, they should be viable enough to sustain those remaining populations at a higher level of living (Conference transcript, 1998).

Regarding the question of moving communities, moving breadwinners from households or destroying the social fabric of families and communities is not intended in any way. Consideration of moving communities refers to people living in non-arable and non-ranching areas, who are there because they have no option. Moving people from such areas is intended to bring additional economic benefits for such families who cannot make a living in such marginal environments (Conference transcript, 1998).

Regarding community area extension, the identification of resettlement land close to communal areas is intended to cause least disturbance to communities who would not want to move away from their current location. Some re-organisation of communal areas is necessary to rationalise land use and settlement patterns (Conference transcript, 1998).

Conclusion

The key issue that has emerged from the views presented here (at the SOAS conference) is that there is in Zimbabwe a well-intentioned and not racially motivated programme that aims to address the critical issues of a national land reform programme. That programme follows the due process as enunciated within the Zimbabwean constitution and the enabling legislation in the form of the Land Acquisition Act. It seeks in the short-term, to achieve the objective of facilitating access to land, in the medium-term, to achieve efficiency gains in utilisation of the land, and in the long-term, to create a broad-base on which long-term economic growth can be predicated (Conference transcript, 1997). It is hoped that financial backing can be secured to ensure its speedy and effective implementation.

References

Conference transcript (1998), Transcript of the Proceedings of the 1998 SOAS Conference: *Land Reform in Zimbabwe: The Way Forward*, 11 March, 1998, Centre of African Studies, School of Oriental and African Studies, University of London.

Government of Zimbabwe (1991), *Land Aquisition Act*, Government Printers, Harare.

Surveyor-General (1984), *Zimbabwe: Natural Regions and Farming Areas.* 1:1,000,000, Department of the Surveyor-General, Harare.

Takavarasha, T. (1998), 'Land Policy and the Land Reform Programme', paper presented at the March 1998 SOAS Conference: *Land Reform in Zimbabwe: The Way Forward*, by Dr. T Takavarasha, Permanent Secretary, Ministry of Lands and Agriculture, Government of Zimbabwe.

Zimbabwe High Commission (1997), *Press Release*, London, December 1997.

6 Theory into Practice: Perspectives on Land Reform of the Farmers' Unions of Zimbabwe

DR T.A.S. BOWYER-BOWER

Introduction

This chapter analyses views of the Zimbabwe's farmers' unions regarding Zimbabwean land reform in general, and the acquisition and settlement plans stated by the Zimbabwe Government in November 1997 in particular. Information is drawn from papers produced and presentations made by each union to the March 1998 SOAS conference 'Land Reform in Zimbabwe: The Way Forward' (Hasluk, 1998; Moyo, 1998; and Zhou, 1998), and from views stated during the conference plenary discussions. The three farmers' unions in Zimbabwe are the Zimbabwe Farmers' Union (ZFU), the Commercial Farmers' Union (CFU), and the Indigenous Commercial Farmers' Union (ICFU). The emphasis of each union's views largely reflects their respective mandates. Each union represents a different target group of farmers, and although established on somewhat contrasting ideologies, their objectives, both practical and ethical, show considerable consensus. The main concerns of these unions are analysed below.

Background to the Farmers' Unions

Zimbabwe Farmers' Union (ZFU)

The Zimbabwe Farmers' Union exists to represent the interests of smallholder farmers in Zimbabwe. It was formed in 1991 from the amalgamation of the National Farmers' Association of Zimbabwe (NFAZ), (the main focus of which was the peasant farmers largely in communal areas), and the Zimbabwe National Farmers' Union (ZNFU), established in

1939, to represent the small-scale commercial farmers in Zimbabwe. Current membership of the ZFU is some 72,000 members. The union stands to represent farmers on communal land, in resettlement areas, as well as small-scale commercial farms. The mission of the ZFU is stated as being:

> The advancment and protection of the interests of the smallholder farmers and the promotion and development of a viable agricultural industry (Zhou, 1998).

Commercial Farmers' Union (CFU)

The Commercial Farmers Union is the original union representing the concerns of Zimbabwe's large-scale commercial farming sector since before independence. The CFU was established in 1942, and united the various commercial unions and associations in existence at that time. It was funded by the compulsory payment of farmers' licence fees and producers levies controlled under the Farmers' Licensing and Levy Act, which was reviewed in 1995, and after which membership has been voluntary. Ever since its inception, the CFU has operated officially as the organization representing commercial farmers' interests. The CFU today has some 4000 members (CFU, 1997). The principal objective of the CFU is stated as being:

> to represent, protect and advance the interests of its members and to further the development of an economically viable and sustainable agricultural industry (CFU, 1997).

A key aim is to provide communication channels within the union at all levels, between the union's farmer members and government and statutory boards, and other organisations connected with the agricultural industry.

Indigenous Commercial Farmers' Union (ICFU)

The Indigenous Commercial Farmers' Union was first registered in 1996, and grew out of an association for indigenous commercial farmers formed in 1990, who were concerned that unfair practice and protection existed in Zimbabwe's large-scale commercial sector. It has some 1000 members today. The founder farmers entered commercial farming after Zimbabwean independence in 1980 and responded to what they saw as unfair laws and circumstances, many of which first arose during the colonial period, posing unique problems, which they felt were not addressed by the other existing unions. Examples are deficiency in financial and technical skills, poor access to financial facilities due to a lack of credit history, as well as discriminatory laws, rules and regulations of the financial institutions. The main focus of the ICFU is assistance to entry-level indigenous commercial farmers, advising new farmers on strategies to overcome constraints through, for example, skills training, preparation of business plans and advocacy (Moyo, 1998). The ICFU's mission statement is as follows:

> Being a lawful organisation which respects merit and equal opportunity in the practice of commercial farming in Zimbabwe, ICFU will use, among other lawful strategies, affirmative action concepts to redress unfair practices, currently militating against new entrants into commercial farming. The main objective of the organisation is to improve the productivity of member farmers through the provision of support services and framework for the attainment of a conducive operating environment.

The role of the union in so doing, is to organise indigenous farmers into a collective group in order to articulate the group's interests to the society at large (Moyo, 1998).

Common Objectives of Zimbabwe's Farmers' Unions

While each union has its own set of objectives arising from its particular social and historical context, there are many aspects of land reform on which they all concur. This section examines these areas of consensus.

The Necessity of Land Reform

All three unions agree that land reform is a pressing issue in Zimbabwe and that plans must be implemented to bring about a more equitable land distribution. All agree that any land reform strategy must contribute to the following objectives:

- indigenisation / equitable land distribution;
- alleviation of poverty;
- alleviation of pressure on communal areas;
- economic revitalisation of communal areas;
- increasing agricultural efficiency and production;
- increasing food security.

All three unions consider that land reform is germane to the above objectives, and that land reform must be carried out in a way that best contributes to these.

All three unions agree on some conditions of successful land reform. Particularly, they agree on three major issues. One concerns the nature of tenure and another concerns the groups and individuals who should be resettled. The third concerns the political process of land reform implementation. These areas of consensus will be examined in turn.

Land Tenure Arrangements

The ZFU, CFU and ICFU all concur on the importance of title deed for achieving the objectives outlined in the last section. They arrive at this conclusion through various arguments.

The Zimbabwe Farmers' Union considers that a secure tenure system is necessary to attract private sector investment. They advocate maintaining the current tenure system such that the collateral security base is also maintained. They do not regard an extension of the communal pattern of settlements as a viable option and consider that the key to the development of communal area agriculture is security of tenure.

The Commercial Farmers' Union considers the institution of property to be fundamental to the whole social order. They consider ownership of property to be a statutory legal concept, and security of tenure to be germane to contemporary development. Hence, they regard tenure law concerning commercial land in Zimbabwe, which enshrines ownership of title deed, as best suited to achieving the objectives of land reform. In contrast, they consider that the law currently applying in communal areas does not promote the well-founded objectives of the Land Tenure Commission (1994). In addition, they are concerned that it is enshrined in Zimbabwe's constitution, that customary law can prevail over the statutory law on which current commercial property rights are based (Chapters 13 and 14 of this text explore these issues in more detail).

The Indigenous Commercial Farmers' Union advocate a review of the land tenure system with the aim of creating and restoring confidence in all sectors of farming. They consider emphasis on freehold title and a lease/tenant system backed by strong guarantees of security, to be fundamental to successful land reform. (For further consideration of land tenure issues refer also to the variety of land tenure options in the various models of land reform proposed in Chapter 16).

Who Should be Resettled?

There is a strong consensus between the farmers' unions in that resettlement must involve experienced and new farmers who can demonstrate the ability to run commercial farms. The ZFU considers that only the best smallholder farmers should find their way onto resettlement land. They indicate that there is evidence to suggest that the top 25 per cent of smallholder farmers can produce at levels above those being achieved by average large-scale farmers. They emphasise that small-scale agriculture currently dominates sectors such as maize and cotton (Figure 6.1) despite constraints on access to land, limited finance and location in poor agroclimatic zones. The ZFU also argues that small-scale agriculture offers greater potential for employment generation than large-scale agriculture (Table 6.1). Hence, the resettlement scheme should offer opportunities for self-employment to competent farmers, in order to maintain and improve productivity.

The ICFU consider that those resettled should consist of competent farmers, able to develop both large-scale and small-scale agriculture. These should be drawn from a number of sources, including the small-scale sector, new graduates and businessmen, retired civil servants and war veterans. All these prospective farmers should be able to demonstrate their abilities and should receive training, particularly apprenticeship, prior to resettlement.

The CFU concurs with the premise that resettlement based on competence, rather than political or social motive, is key to maintaining and enhancing productivity.

The Land Reform Process

All the farmers' unions agree that the success of the Land Reform Programme will rest on the degree of transparency achieved by the government in carrying out its objectives. Involvement of all stakeholders is considered germane to the planning and implementation process. This is important so that indigenous Zimbabweans and international donors and investors alike, can observe that land reform is carried out in a way that best contributes to the objectives outlined at the beginning of this paper. Given the vulnerable process of land reform, the ICFU advocate precautions to ensure that those with ill-gotten wealth or corrupt intentions do not gain access to resettlement land.

Figure 6.1 Market domination of maize and cotton production by smallholders

Source: Adapted from Zhou, 1998

Table 6.1 Numbers employed on large-scale commercial farms compared with smaller-scale farming (i.e. small-scale commercial farms and resettlement areas)

	Owners, Plotholders Lesees actively engaged in farming		Employees		Total for farm type		Crop area		Hectares per Person (ha/p) employed	
	1990	*1994*	*1990*	*1994*	*1990*	*1994*	*1990*	*1994*	*1990*	*1994*
LSCF[1]	5,704	6,496	257,468	310,944	263,172	317,440	494,981	498,572	1.88	1.57
SSCF[2]	229,544	241,318	21,802	22,483	251,346	263,801	70,297	74,890	0.28	0.28
RA[3]	768,270	861,330	9,525	9,588	777,795	870,918	142,055	158,711	0.18	0.18
Total 'employed'[4]	1,003,518	1,109,144	288,795	343,015	1,292,313	1,452,159	707,333	732,173	0.55	0.5

1 Large-scale commercial farms
2 Small-scale commercial farms
3 Resettlement areas
4 Total 'employed' for type of employment.

Source: Central Statistics Office, adapted from Zhou, 1998

The Land Acquisition Issue

All unions also articulate their support for the means of acquisition and resettlement to be carried out within Zimbabwe's legal framework. Furthermore, they share concerns that previous resettlement has been undermined by putting political considerations ahead of economic ones. In order to ensure that the objectives of land reform are fully realised, each union emphasises its own role in the planning and mobilisation process.

With regard to analysis, planning and implementation, each union emphasises different aspects of the process and offers various solutions and procedures. Provided that the land reform process is transparent and that the farmers' unions are satisfied with their level of representation, the current position of each union does not constitute any major contradictions to government policy, except over the issue of land acquisition. Hence, the process for land acquisition is the main aspect of contention between the unions, with the ZFU and ICFU generally aligned with current government plans, and the CFU advocating a free market mechanism.

Both the ZFU and ICFU support the government plans for land acquisition, including legal compulsory acquisition should land owners not agree to sell. They contend that land reform is long overdue and that further delays will have dangerous implications for social stability in Zimbabwe. They argue that the process has been held up by the unwillingness of the commercial sector to offer adequate quality land for resettlement. The identification of land for resettlement should therefore be coordinated by government and carried out by technical planners and representatives of the ZFU and ICFU. Both unions accept the need to ensure that fair compensation for land is received.

The CFU does not support compulsory acquisition of land for commercial purposes, arguing that such a policy would seriously undermine perceptions of tenure security. While compulsory acquisition is legal under the 1992 Land Acquisition Act, they claim that investors and donors favour market-led distribution of land and would be willing to fund preferential loan facilities to enable this process of redistribution. Furthermore, if the government were to agree the CFU's proposals, they will be willing to mobilise their own institution and other supportive institutions to implement the land reform process.

The CFU hold that compensation of current landowners would not constitute a benefit from colonial occupation since all land was acquired lawfully. Furthermore, they argue, 70 per cent of the 1471 farms implicated in current acquisition plans (by having been gazetted) were purchased after Zimbabwean independence.

The ZFU and ICFU support proper compensation, and as such, all three unions are in agreement about the requirement to pay fair compensation. What constitutes fair compensation, however, is still an issue of debate. The ICFU argue that previously, departing farmers have removed much of the infrastructure that settlers had expected to still be in place. The CFU argue that land and infrastructure form only part of the balance sheets of farm financial statements. They contend that the manner in which compensation is made is obviously of great concern to the owners listed in the government's current acquisition plans, and crucial to the sustainability and success of the Land Reform Programme.

Specific Perspectives of each Union

Since each institution has arisen in relatively different contexts and since they represent different parties, it is unsurprising that they also emphasise different concerns. The following section will examine these concerns in turn.

Perspectives of the Zimbabwe Farmers' Union

The ZFU fully support the Government's acquisition plans, contending that the land reform issue is well overdue and cannot continue to be postponed. They believe that the speedy conclusion of the land reform process is necessary politically, economically and socially, and is integral to continued peace and stability in Zimbabwe.

The ZFU accepts the government target of acquiring 5 million hectares of land. However, the final target is to resettle 8.3 million hectares. The ZFU hold that land should be identified such that when the final figure of 8.3 million hectares is realised, the resettlement sub sector should have the same land distribution pattern in the various agro-ecological zones as the large-scale commercial sector.

The ZFU emphasise the ability of smallholder farmers to produce at levels comparable to the large-scale commercial sector. They support the government principle of emphasising competency as a basis for settler selection but point out that social and political considerations have tended to take precedence over economic considerations. Settler selection should be conducted through an appropriately constituted panel, headed by the Ministry of Lands and Agriculture and involving appropriate institutions.

The ZFU consider funding of the resettlement scheme to be a vital prerequisite of success, along with selection of competent farmers and secure land tenure arrangements. Present settlers have met serious difficulties in financing their farm operations due to an unfavourable economic climate, high interest rates and lack of secure tenure for collateral. The ZFU advocate the establishment of a resettlement fund to provide a 'financial window' for new settlers. Funds for this and other provisions could be drawn not only from government and donors, but also from the following strategies:

- An annual fee, pegged at reasonable levels, to be paid by all settlers. This could be in the form of a lease fee and levied only on the condition that land is subdivided into viable self-contained units;
- The Development Levy that is currently being collected from employees by the government;
- Land Tax on all commercial land.

Finally, because the resettlement scheme will only affect a limited number of people, conditions in communal areas must continue to be improved, in order to facilitate the commercialisation of agriculture. The key impediments to this process are tenure insecurity, inadequate communications and lack of irrigation infrastructure in the communal areas. These must be addressed by increased investment in order to widen the income base and demand for goods and services. This, in turn, will stimulate increased industrial and economic growth in Zimbabwe.

Perspectives of the Commercial Farmers' Union

The CFU have developed a policy on land acquisition and redistribution within the context of ongoing liberalisation of the Zimbabwean economy. They emphasise the development of sustainable agriculture by means of improving agricultural efficiency and increasing farm incomes. They consider that the acquisition of land must be judged in terms of its contribution to economic growth, employment growth, food security, balanced development and trade facilitation.

The CFU emphasise that land acquisition should be conducted lawfully and that fair compensation should be paid, in line with the 1992 Land Acquisition Act. They contend that since farmers carry heavy debt burdens, inadequate compensation would undermine the finance industry. Confiscation of land will imply a capital loss for both the farmer and the financial institution supporting the farmer. If financial institutions incur these losses they are unlikely to be willing to lend to new settlers, if indeed they survive. Hence, the CFU's approach to land reform is that acquisition should be on sound economic and financial grounds which necessarily entails fair compensation. They also highlight the support of international institutions, such as the International Monetary Fund (IMF), for fair payment of compensation within the legal framework of Zimbabwe.

To this end, the CFU propose that resettlement for commercial reasons should be on the basis of identified land being offered for sale to the indigenous commercial sector, who would be supported by preferential loan facilities. Hence, land for commercial resettlement should be taken out of the government's Land Acquisition Programme and the Land Acquisition Act applied only to derelict, vacant and under-utilised land, land owned by foreigners and multiple owners, and land required for social and political reasons.

The CFU paper (Hasluk, 1998) declares that they are in a position to offer 1.5 million hectares of land immediately, should the Zimbabwean government agree to the CFU's approach. In addition, they emphasise their own role in sourcing and mobilising funding and technical and legal expertise. They advocate a strategic approach to land reform including detailed analysis and assessment of issues and options, involvement of all stakeholders and representative institutions, coordination and involvement of donor agencies, and strong communication of the positive impact of land reform through a well-planned marketing campaign. In conclusion, they emphasise that security of tenure must be demonstrated throughout the land reform process if donors, investors and farmers themselves are to be assured that ownership is at the heart of Zimbabwe's future land policy.

Perspectives of the Indigenous Commercial Farmers' Union

The ICFU, being set up to champion the misgivings of its members regarding unfair practice, is very concerned with redressing current inequities in land distribution and access to finance and markets for Zimbabwe's indigenous farmers. They fully support the government plans for lawful acquisition of land and consider that the government should play a 'coordinating and facilitating' role in a process that should involve all interest groups. Moreover, the land reform process should not be considered as an end in itself, but should be taken in conjunction with the overall national vision on development.

The ICFU emphasise the need to plan for tailor-made support services for resettled farmers, based on the same principles which were so successfully undertaken to support white settlers in colonial times. The ICFU consider that the historical experience of apprenticeship schemes should be taken into account in land reform planning. Prior to independence, a system of apprenticeship trained young farmers for 2-5 years before they moved to farms of their own. The ICFU conclude that the promotion of *good neighbourliness*, as supported at national level, was instrumental to the success of the planned settlement programmes of the Rhodesian government prior to independence. The ICFU emphasise their role in bringing about genuine cooperation between indigenous farmers and their ability to provide much-needed information and training in a number of key areas, including technical expertise and business management.

The ICFU are very concerned with lack of access to finance and the extremely high interest rates prevailing in Zimbabwe today. If productivity is to be maintained, resettled farmers will require access to finance based on ownership of title, and to loans with more favourable interest rates. In the light of this, a variety of financial packages should be available, including development loans for farm machinery and livestock equipment.

Other aspects of the ICFU's conceptual framework include the clear demarcation between 'subsistence-level resettlement models' (mainly for the alleviation of land pressure in communal areas) and 'commercial resettlement models' (mainly for the production of national strategic agricultural products on a commercial basis). The strategy for the latter sector should include cost recovery mechanisms, commensurate with the long-term productivity of this sector. Similarly the ICFU are concerned with the rationalisation of land size in terms of economically viable units in the various agroclimatic zones and the development of basic infrastructure to support these as viable land units. Properties to be offered for sale should be fully described in a prospectus, listing all fixed assets and costs. Production targets based on land capability classification should be set and support given to ensure those targets are achievable. Prospective settlers should be required to formulate and submit farm plans and their performance should be monitored and assisted on a continual basis.

Conclusion

The amount of land available for resettlement, its price, arrangements for land ownership, the consideration of who should be resettled, or the means of land acquisition, are not the only issues that need to be resolved for land reform in Zimbabwe to be a success. All three unions also spoke of the important need for a comprehensive and open system of management for the subsequent resettlement process, and adequate support for the broad needs of the subsequent production systems on the resettled land, if they are to be sustainable in the longer term. Without this, land reform, with all its pains and fears, becomes an end in itself rather than a means to an end, and could instead detract from, rather than contribute positively to, the stated intended goals of the land reform process. It is important that the financing of land reform schemes fully reflects this holistic need, and not just the land acquisition stage of the land reform process, which is but one part of the whole, and cannot in itself guarantee success. It was widely agreed that the continued positive collaboration of all three unions together will be essential in achieving this goal, and is to be applauded and encouraged.

References

CFU (1997), *Commercial Farmers' Union Information Brochure*, Commercial Farmers' Union, Zimbabwe.

Hasluk, D. (1998), *An analysis and some proposals on the Zimbabwe Government Land Acquisition and Land Reform Programme,* CFU paper presented to the March 1998 SOAS conference: 'Land Reform in Zimbabwe: The Way Ahead', by Mr David Hasluk, Director, Commercial Farmers' Union, Zimbabwe.

Land Tenure Commission (1994), *Report of the Commission of Inquiry into Appropriate Land Tenure Systems*, Government printers, Harare.

Moyo, N. (1998), *Land redistribution, use and management in Zimbabwe: A concept paper*, ICFU paper presented to the March 1998 SOAS conference: 'Land Reform in Zimbabwe: The Way Ahead', by Mr Nokwazi Moyo, Executive Director, Indigenous Commercial Farmers Union, Zimbabwe.

ZFU (1997), *Zimbabwe Farmers' Union Constitution,* Zimbabwe Farmers Union, Zimbabwe.

Zhou, E. (1998), *Smallholder farmers' viewpoints on land acquisition and distribution*; ZFU paper presented to the March 1998 SOAS conference: 'Land Reform in Zimbabwe: The Way Ahead', by Mr Emerson Zhou, Deputy Director, Zimbabwe Farmers Union, Zimbabwe.

7 The Political Economy of Land Redistribution in the 1990s

PROFESSOR SAM MOYO

Introduction: Economic Basis of Land Reform

Over six million people live in Zimbabwe's marginal rural lands, restricted from access to the bulk of the nation's natural resources. Inequitable access to these resources means that 4,500 mainly white, large-scale farmers dominate Zimbabwe's agrarian economy. These imbalances dramatically skew Zimbabwe's income distribution structure and reflect an unchanged legacy of colonial rule. The growth of poverty, unemployment and income disparities due to the under-utilisation of Zimbabwe's land and natural resources, is the main factor which fuels today's land question.

The key issue facing Zimbabwe's land reform policy is how to balance the control and access to land by redistributing land from large-scale landholders who underutilise their land to new small and medium-scale users. The objectives of this analysis is, therefore, to evaluate the political economy of Zimbabwe's emerging land policy.

Perceived Economic Impacts of Land Transfer

Three extremely simplistic frameworks of analysing the costs and benefits of acquiring the 1,471 gazetted farms have been proffered, mainly by private sector and media interests. First, a general political and economic approach, of listing a series of individual agro-economic and political consequences of land transfers, is offered. Second, a commodity output loss approach (cropped area, volumes and values) has been attempted, using weak and vague quantitative methods. Finally, a macro-economic framework using broad quantitative estimates is proffered, again based upon weak information sources. Most assessments tend to be static and focused mostly on the psychological effects that tampering with property relations may have on markets or investors. Few assess the economic, social and political benefits which could be realised from the new farmers who gain access to land, whether or not these are for own consumption.

73

The debate has thus been only cost-oriented and not benefit-oriented. The main commodity losses anticipated focus on tobacco, cotton, horticulture, sugar and maize, while losses among field crops such as wheat and soybeans are rarely cited. In fact, there is little expectation of major losses in food outputs or of increased food imports as a result of the land acquisition. The major fear is of losses of exports.

Tobacco is expected to lose 50 per cent output since purportedly 700 tobacco farmers' lands were gazetted for acquisition in November 1997. Tobacco alone accounts for 40 per cent of Zimbabwe's total exports, such that tobacco production losses from land acquisition is predicted to result in a 20 per cent loss of foreign currency by 1999, as small farmers are not expected to master specialised skills.

The value of outputs for cotton, maize and horticulture are all expected to decline by 50 per cent by 1999, according to the CFU (Commercial Farmers' Union). But losses of maize and cotton are quite controversial since small farmers already produce 65 per cent of the national totals (Chapter 6 of this text refers to this further). Moreover, the land area required for horticulture is quite small such that producers can easily access such land. Contradictory reports arise around whether new planned investments in sugar production will be lost.

One key problem with the above analyses is their tendency not to differentiate the sources and effects of the problems that could arise from land acquisition. The varied effects of changes in rainfall, technology and irrigation on commodity production are for instance hardly considered. Moreover, the loss of agricultural output is not entirely novel to Zimbabwe, given that the agricultural sector has to regularly adapt to droughts. The large-scale commercial farms (LSCF) have been the most successful in adapting to such shocks of weather and shifting land uses.

Only 46 per cent of the 4,500 LSCF farmers have irrigation facilities to stabilise crop production, although their numbers grew to this level mainly since the mid-1980s, due to subsidised state credit. Some estimates indicate that only 180,000 hectares (30 per cent) of the cropped area is irrigated fully (FAO, 1990), while altogether 46 per cent of the LSCF farms may have access to full and supplementary watering using varied water sources to stabilise water availability from rainfall fluctuations. Farmers in Natural Region II alone, are responsible for 56 per cent of Zimbabwe's total irrigated land, while another 30 per cent of the irrigated area is found in Natural Region V, among a few multinational LSCF holdings which were not listed for acquisition. Thus inconsistent representation of output trends is common in Zimbabwe even though the frequency of droughts requires more precise estimations of the scale and cause of crop losses.

More LSCF maize is being used on-farm for livestock feeds, raising other questions about the efficiencies of Zimbabwe's livestock production system, especially of allocating land and maize produced in Natural Regions I and II. There has also been a switch among some LSCF farmers' land use in prime and poorer natural regions from commercial beef production towards wildlife management (Moyo, 1998a).

Actual Implications of Acquiring the 1,471 Farms

The policy discussion on the economic implication of land acquisition has become misleading because the scale of land targeted has not been adequately conceptualised. The government policy sought to acquire farms which it referred to as 'too large'. No clear-cut policy definition of the farm size variable is available besides a general 1,500 hectares as the threshold size for viability in Natural Regions I and II.

Farm Sizes and Production Potential

A careful analysis of the relationship between farm numbers and land area identified shows that a handful of very large farms dominate the distribution of land-ownership. There is an inverse relationship between farm numbers and area amongst farms in the LSCF. There tends to be a larger number of the LSCF sector farms found amongst relatively smaller-sized farms (area-wise), while a few farms occupy extremely large expanses of land area.

Examining the farm size classifications, we found that two categories of farm size accounted for significantly large areas of the land identified. Farms between 1,500 and 2,999 hectares amounted to 18 per cent of the farms and accounted for 14 per cent of the area, while those between 5,000 up to 14,999 hectares accounted for 8 per cent of the farms and 25 per cent of the area. Yet 66 per cent of the identified land area consisted of only 222 farms all being over 3,000 hectares in size, and some of them reaching up to 350,000 hectares in size.

Thus, a few large farms (which are mainly underutilised) accounted for the bulk of the identified area, suggesting that the pace of acquisition could be managed according to the farm number and area relationship, and that it may be less cost-effective to designate many small farms.

Agrclimatic Potential or Features

The quality of land being acquired in relation to scale is, therefore, a critical economic policy measure of the land policy's efficacy. In this regard, the nature of agroclimatic potential of the identified lands was a crucial variable in defining the future potential of resettlement schemes, as well as in terms of gauging the broad potential effects of acquisition on current farm production activities.

Our evidence shows that over 62 per cent of the identified land was located in Natural Regions IV and V and up to 80 per cent was located between Natural Regions III and V. Thus, only about 704 of the identified farms were in Natural Regions I and II. These amounted to below 20 per cent of the area identified. It is these farms which need to be examined closely as their high potential, assuming it was being used, could be lost by acquisition *per se*, before we look at resettlement output potentials.

LSCF Land Tenure and Ownership Structure

It is possible to identify six land ownership categories among farms listed for acquisition. These are:

- Companies (58%)
- Individuals (24%)
- Government parastatals (less than 2.3%)
- Churches and NGOs (1.6%)
- Multinationals (13%)
- Two others are owned by the National Railways of Zimbabwe and the Cold Storage Commission.

Multiple Farm Ownership Patterns

The issue of multiple ownership of farms can be used to determine farm management and land utilisation efficiencies. The data excludes a large number of multiply-owned farms belonging to owners who do not appear on the identification list. It does not tell us about additional farms that the listed farmers may own. This is to say there is much more multiply-owned farmland which did not necessarily appear on the current identified list.

The land tenure evidence shows a diverse and differentiated structure of landholding and land-use among Zimbabwe's white population. On the one hand, we have a few white-dominated large companies owning the greater part of Zimbabwe's commercial farmlands, which are relatively underutilised. On the other hand, we have a combination of 1,000 white-owned family farms and family-based companies each owning farms that are relatively small, as well as a few large companies with relatively large farm areas who use their land relatively adequately. The debate on compulsory land acquisition should, therefore, be more nuanced and qualified in its assessment of the land targeted for acquisition.

The Social Implications of Land Transfer

Uneven Acquisition Pattern

Land acquisition has, to date, been unevenly focused on five provinces. Mashonaland Central, Matabeleland North and Mashonaland East have been areas least targeted for land acquisition.

Of the country's 55 Rural District Councils, 44 districts had land identified. This gives an average of five districts affected in each province. But only 25, approximately half of the country's districts, accounted for 83 per cent of farms and 90 per cent of the total areas identified. Just 10 districts accounted for 63 per cent of the area identified, while the other 10 accounted for less than 3 per cent of the area. Not surprisingly, some areas with large extensive farm areas and high levels of underused land, ranked highest among identified areas. Prime lands with historically low levels of land utilisation were spared intensive land acquisition.

Communal Area Borders

One of the justifications for listing farms was their proximity to communal areas (CAs) facing severe land pressure and landlessness. Originally this criterion was intended to reduce the costs of moving settlers, to expand the land available to local communities or provide 'elbow room' (Cliffe, 1988), and avoid bringing in 'strangers'. While these reasons bear on the efficiency of resettlement, the chief rationale for land reform accepted by most stakeholders has more to do with the principle of disallowing the underutilisation of land.

Proximity to communal areas is a problematic concept, since all commercial farms are brought closer to these areas once the farms between them and the nearest communal area are identified. Government officials recognise that this poses a problem of shifting boundaries and can be used during one phase of land acquisition only.

Zimbabwe's Landed Gentry

Using the cut-off point of over 10,000 hectares, 66 landowners occupying 2,108,972 hectares were identified as critical to negotiations for land transfer. Among these we found eight individuals among the targeted farms who together owned 13 farms occupying 158,531 hectares, of which 29 per cent of the area was owned as multiple farms. Thus, only five persons who owned over 10,000 hectares owned these as single farms amounting to 112,537 hectares. Multiple farm ownership is thus a decided feature of Zimbabwe's landed gentry, whether company or individually owned.

There were black companies which held five farms of close to 17,000 hectares in our area-based definition of the landed gentry. Casual empirical observation suggests that there could be up to 10 more black owners of over 10,000 hectares among those whose land was not designated.

Only 20 wholly black-owned landholding companies were among those targeted for land acquisition. Five of these farms had multiple farms, while the 28 farms owned by the 20 firms occupied about 2 per cent of the entire company owned land. More blacks own land in their individual title.

Sex Distribution of Land Ownership

Over 72 per cent of the registered land owners were male, about 23 per cent of the farms were jointly owned and less than 5 per cent were owned by women. Less than 29 per cent of the directors of these farms were women. Overall, less than 6 per cent black women owned land. White women were mostly registered as joint husband and wife owners. Thus land transfer negotiations will mainly have to be held with white males.

Absentee Landlords

Nationality can be a crude measure of 'absentee' ownership. In turn, absenteeism is often seen as a reflection of limited commitment to farming. Absentee owners rely on farm managers supervised from afar. In Zimbabwe, absenteeism has a particular significance because a significant number of farm managers are actually blacks whose skills are not appropriately recognised. Black managers are often classified as 'supervisors' and 'semi-skilled' and receive lower wages than their jobs warrant. But because about 40 per cent of farm management and technical skills are indigenous, according to survey evidence, it could be argued that since LSCF are essentially black managed, the transfer of land ownership towards such farmers will not have negative effects.

Race, Nationality and Indigenisation

Achieving an equitable balance in the racial and national origins of land ownership has been a key political objective of Zimbabwe's Land Reform Programme. Just under 250 black indigenous LSCF owners, comprising about 17 per cent of the listed owners, had their farms targeted by the provincial land identification committees. These were mainly smaller sized farms of less than 3,000 hectares. But, together with the larger black-owned farms, they amounted to about 20 per cent of the total area identified. However, less than 10 per cent of the total LSCF sector is, in fact, owned by black indigenous persons. Thus, the indigenously owned land identified for purchase may be regarded as having been disproportionately targeted.

The view now emerging within the government is that the acquisition of black-owned farms at this stage defeats the objective of indigenisation. But some 'indigenously-owned' farms may actually have deserved to be targeted because their farms are extremely large and comprise parts of huge multiple estates.

Ethno-Regionalism and Land Reform

It is puzzling that about 100 small farms located in Natural Regions II and III and owned by indigenous persons should also be targeted, since these have the potential to be productively used and reflect a desirable move towards medium-scale farms.

The manner in which elites, such as government supporters, will gain access to land is a matter of public speculation. Fear of discrimination in land redistribution and the wider indigenisation policy process heavily shadows the decision-makers in charge of land reform.

Moving Forward with Land Reform Policy-Making

Credibility of the Policy Process

The government has tended to rationalise land acquisition and redistribution on the basis of historical grievances and political demands, which in their own right are legitimate, over and above the actually existing valid economic rationale for land reform. Indeed, some public statements on land acquisition have tended to imply that there are no sound economic objectives underlying the policy.

Because of the poor supply of objective information by government and private sector stakeholders alike, rumour and subjective interest rather than facts have led strategic policy thinking. Thus, the credibility of land reform is affected by the absence of a transparent plan and policy dialogue process, which are necessary to counteract attempts to sensationalise the perceived relative costs and benefits of land acquisition.

Economic Rationale and Goals of Land Reform

Inequitable access to resources means that the structure of the Zimbabwean economy continues to undermine the growth of rural incomes and restricts the expansion of domestic markets and industrial development. The key objective of the land reform policy must be to enhance economic expansion and diversification through expanding inputs from a more efficient and rational structure of farming and land resource utilisation.

Such a land policy cannot depend on the narrow consumption interests and limited effective demand of minority elite groups. The transfer of land should thus not be focused on simplistic and emotive processes of off-loading underused marginal lands to the poor and small farmers as a means of extracting new forms of financial and political capital. The issue is to evolve a comprehensive socio-economic development strategy in which land reform is central.

Land redistribution must be directed at an appropriate range of beneficiaries who will have to follow land use, outputs, income and employment targets established in relation to clear goals, and the quality of land and related resources allocated to them. The land redistribution policy and plans must be adequately detailed in terms of numbers of different types of beneficiaries to be allocated different types of lands. Those limited plans as exist need to be more adequately coordinated among the relevant implementing agencies. The public needs to be fully informed about plans and assured about the fairness of the redistribution. A Land Commission could be tasked with policy formulation and implementation. It should ensure that the land reform plan should be phased over the next five to ten years. During the first five years, a phased acquisition of the targeted five million hectares should be undertaken beginning with half of this land between 1998 and 1999 and the other half between 2000 and 2001.

References

Alexander, J. (1995), 'The Unsettled Land: The Politics of Land Distribution in Matabeleland, 1980-1990', *Journal of Southern African Studies*, 17, 4.

Cliffe, L (1988), 'Zimbabwe's Agricultural Success and Food Security', *Review of African Political Economy*, 43.

The Economist, *Populism Awry*, 24 January, 1998

FAO (1990), *Constitutional Amendment of the Land Policy*, Hansard/Government Gazette.

Financial Gazette (1998), 29 January, 1998.

Government of Zimbabwe (1992), *Land Acquisition Act*. Harare: Government Printers, Harare.

Government of Zimbabwe (1996), *Land Redistribution and Resettlement*, Policy paper, Government Printers, Harare.

The Herald, (1996), November 1996.

Land Tenure Commission (LTC) (1994), *Report of the Commission of Inquiry into Appropriate Land Tenure Systems*, Government Printers, Harare.

Moyo, S. (1986), 'The Land Question', in I Mandaza, (ed.) *Zimbabwe: The Political Economy of Transition, 1980-1986*, CODESRIA, Dakar.

Moyo, S. (1995), *The Land Question in Zimbabwe*. SAPES Books, Harare.

Moyo, S. (1997), 'Land Reform or Political Posturing?', *The Mirror* Vol. 1 (1), 1-7 December 1997.

Moyo, S. (1998a), 'Land Acquisition will not Lower Productivity', *The Mirror* Vol. 1 (7), 19 - 25 January, 1998.

Moyo, S. (1998b), 'Land Acquisition: The Social and Political Context', *The Mirror* Vol. 1 (8), 26 January - 1 February, 1998.

Moyo, S. (1998c), 'The Framework for the Design of Land Reform', *The Mirror* Vol. 1 (9), 2-5 February, 1998.

Moyo, S. (1998d), *The Land Acquisition Process in Zimbabwe-1997*, SAPES Policy Monograph Series, No.1.

ODA (1996), *ODA Land Appraisal Mission Report*, British Development Division in Southern Africa, Harare.

Palmer, R. (1996), 'Review of Sam Moyo: The Land Question in Zimbabwe', *Development in Practice*, Vol. 6 (4), November 1996.

Potts, D. and Mutambirwa, C. (1997), 'The Government Must Not Dictate Rural-urban Migrants' Perceptions of Zimbabwe's Land Resettlement Programme', *Review of African Political Economy*, 74, pp. 549-566.

Roth, M., Barrows, R., Carter, M. and Kanel D. (1989), 'Land Ownership, Secuirty and Farm Investment: Comment', *American Journal of Agricultural Economics*, 71.

The Sunday Mail (1998), American express article, January –February, 1998.

Weiner, D. (1989), 'Agricultural Restructuring in Zimbabwe and South Africa', *Development and Change,* 20, 3.

World Tribune (1998), *Zimbabwe's White Farmers Offer Redistribution Plan: They would sell land to State and Aid Peasants,* The Associated Press, 24-25 January, 1998.

8 Implications for Poverty of Land Reform in Zimbabwe: Insights from the Findings of the 1995 Poverty Assessment Survey Study

DR T. A. S. BOWYER-BOWER

Introduction

With the change of government in Britain in May 1997 came a change in the objectives of UK spending on international development. The Overseas Development Administration (ODA) under the Minister for Overseas Development (within the Foreign and Commonwealth Office), has become the Department for International Development (DfID), under the Secretary of State for International Development, which is now a cabinet position. The new government has prioritised eliminating poverty as its focus for development spending overseas (Her Majesty's Government, 1997).

While already all agree that the solution to the land reform issue in Zimbabwe is for Zimbabwe to determine, the country will undoubtedly be offered help from a number of allies in the international community. Each of those will have their own interests, priorities and objectives. Allies have to date been requesting that the acquisition and resettlement process be undertaken in a transparent and legal manner that conforms to the constitution of Zimbabwe, and that it should not threaten the economic viability of the nation. A widely acclaimed aim of land reform is to redress the current unequal distribution of land ownership and rights of access to the land resource base of the country in order to redress moral and political wrongs of the past (refer, for example, to Moyo, 1994). As well as focusing on equity, it is essential that the productivity from of the land is also maintained.

A further question being asked is the extent to which the current land reform programme can be assumed to be a means by which poverty in

Zimbabwe will be alleviated and reduced. There are many answers to this question; answers which address direct and indirect links between land distribution, land ownership, and poverty; and which involve implications for both macro-scale systems, such as for national export earning; and more micro-scale systems, such as providing enhanced opportunities for individual or household subsistence, wealth creation, and food security. A conclusion of the Zimbabwe's Ministry of Agriculture's Agriculture Policy Framework for 1995-2020 (Government of Zimbabwe, 1995) asserts that if land reform raises the productivity and incomes of smallholder agriculture in Zimbabwe, it will then be a direct route towards eradicating poverty, hunger, malnutrition and unemployment.

Many of these implications and linkages are touched upon by other papers in this text. Some suggestions are based largely on conjecture, others the result of empirical field-based study. One such field study of relevance here is the 1995 Poverty Assessment Study Survey (PASS, Government of Zimbabwe, 1997). The results of this survey are used here to obtain further insight into likely links between land reform and poverty in Zimbabwe.

Introduction To The 1995 Poverty Assessment Study Survey (PASS)

The PASS was undertaken with multi-national funding, sources of assistance included the United Nations Development Programme (UNDP), the Canadian International Development Agency (CIDA), Danish International Development Agency (DANIDA), the Norwegian Agency for International Development (NORAD), and the International Labour Organisation (ILO). The main report of the survey was released in November 1997 (Government of Zimbabwe, 1997). The survey was government led, but undertaken in close collaboration with non-governmental organisations, the donor community, and civil society.

The survey used three questionnaires, one addressing households, another communities, and the third the homeless. In a total of 809 Enumeration Areas (EAs - those of the 1992 Population Census), during August to November 1995, 19,173 households were interviewed (every fifth household in each EA), 617 communities (the schools and clinics most used in the EAs), and 518 homeless persons. The interviews were undertaken August – November 1995. The survey, as with the results, were stratified not only by province, and by district, but also by land-use sector. In the rural areas, four sectors were distinguished: large-scale commercial farms (LSCF), small-scale commercial farms (SSCF), resettlement areas (RA), and communal areas (CA) (Government of Zimbabwe, 1997).

It is increasingly recognised in international debate that poverty can most usefully be identified not just as an inability to meet nutritional needs and a defined set of non-food expenditures, but also exclusion from access to resources, from knowledge, and from rights (UNDP, 1996; OECD/DAC, 1996; OXFAM, 1996). This moves the definition of poverty more from 'the deprivation of basic human and social needs' to 'a lack of well-being' (DANIDA, 1996). An increasing emphasis is being placed on indicators which take into account these dimensions (for example, the Human Development Index of the UNDP). More conventionally the most straightforward means of measuring poverty, however, focuses on an ability to meet nutritional needs and a defined set of non-food expenditures. The 'wealth' of those responding to a poverty survey is conventionally determined either by measuring household incomes or determining consumption expenditure using survey techniques. Although least accurate (in terms of being commonly underestimated), reported household income is the most easily determined measure of household wealth, and was used in this 1995 PASS survey.

Two poverty levels were distinguished for the survey:

1   the Food Poverty Line (FPL): defined as the amount of income
     required to buy a basket of basic food[1] needed by an average person
     per annum;

2   the Total Consumption Poverty Line (TCPL): defined as the amount of
     income required to purchase the basket of food, and non-food
     (clothing, housing, education, health, transport, etc.), by an average
     person per annum.

Those living below the FPL are categorised as 'very poor', those living above the FPL but below the TCPL are categorised as 'poor', and those living above the TCPL are termed 'non-poor' (Government of Zimbabwe, 1997).

The results of this survey provide a wealth of information on poverty, including the demographic, geographic and sectoral distribution of poverty in Zimbabwe; and insight into the interrelationship between poverty and access to services (education, health, roads, etc.); and poverty and the state of the environment (land degradation, agriculture, etc.). One of the many interesting parts of the study was its exploration of perceptions of poverty and what the poor themselves see as the most effective actions for poverty reduction. Undoubtedly many highly valuable inferences can be made from the findings of this study in answer to a diversity of questions for a number years to come. Some of the results that more directly indicate likely implications for poverty of land reform in Zimbabwe are examined below.

Findings of the 1995 Poverty Assessment Study Survey

Defining Poverty in Zimbabwe

In answer to the question: 'what constitutes poverty in Zimbabwe?', the food poverty line for Zimbabwe was determined as being Z$1,289.81 nationally (Z$1,511.77 in urban areas compared with Z$1,180.49 in rural areas); the total consumption poverty line for Zimbabwe was determined as being Z$2,132.33 nationally (Z$2,554.89 in urban areas compared with Z$1,924.20 in rural areas. See Table 8.1).

Table 8.1 Poverty levels in Zimbabwe (Zim$ per person per annum) - results of the 1995 PAS Survey

	National	*Urban*	*Rural*
Food Poverty Line $	1,289.81	1,511.77	1,180.49
Total Consumption Poverty Line$	2,132.33	2,554.89	1,924.20

Source: Government of Zimbabwe, 1997

The Incidence of Poverty in Zimbabwe Nationally, its Prevalence in Urban compared with Rural Areas, and its Relationship with Gender

This breakdown of the PASS results is presented in Table 8.2a & b. The survey determined that 74 per cent of the population of Zimbabwe is poor. This comprises 57 per cent living below the Food Poverty Line (very poor) and the further 17 per cent living above the Food Poverty Line but below the Total Consumption Poverty Line (poor). It is cautioned that these figures may somewhat overestimate levels of poverty in Zimbabwe at the time of the survey because their measurement has been based on reported household incomes which are commonly underestimated (DANIDA, 1996).

Table 8.2a Poverty incidence nationally, rural compared with urban - results of the 1995 PAS Survey

	National	*Urban*	*Rural*
% very poor	57	28	72
% poor	17	22	14
= % total poor	74	50	86
% non-poor	26	50	14

Source: Government of Zimbabwe, 1997

Table 8.2b Poverty incidence by gender - results of the 1995 PAS Survey

	Males	*Females*	*Male-headed households*	*Female-headed households*
% very poor	55	60	40	57
% poor	17	17	18	15
= % total poor	72	77	58	72
% non-poor	28	23	42	28

Source: Government of Zimbabwe, 1997

Notwithstanding this possible influence on these results, the levels of poverty revealed by this survey are a considerable burden for any one country, and illustrate the urgent need for implementing schemes for poverty alleviation and reduction, if a sound and sustainable development is to be achieved in the country as a whole.

Considerably greater poverty was found in rural areas compared with urban areas (86 per cent of the rural population compared with 50 per cent of the urban population: Table 8.2a). In terms of land reform, this finding illustrates that a more lucrative living has been made from urban-based employment activities compared with rural, and despite the higher cost of living in urban areas than rural areas, the result is a lower incidence of poverty in urban areas. The current schemes of land reform are largely resettling poorer families and individuals from one rural area to another (mostly from communal areas to resettlement areas). In terms of providing the landless with land, if the land that is provided is currently unutilised, and as long as the means for making the resettlement land productive is available, then this land reform will help raise average income levels in rural areas, and help alleviate some of the greater poverty here. If, however, the land acquired for land reform is already the employment base of one population, which is displaced as a result of the land reform going ahead, then depending upon what alternative means of living is available for the displaced populations, levels of rural poverty could worsen. Also if the means for making the resettlement land productive is not available (because of a lack, for example, of the necessary support services), then again poverty levels will not be improved.

This data also suggests that a move from a rural-based income activity to an urban-based income activity is likely to be more lucrative than a move from one rural-based income activity to another, and as such, could continue to be a more attractive option (i.e. more attractive than land reform) for those to whom this option is available. Sadly the current economic climate in Zimbabwe is reducing the availability and security of employment opportunities, making this option increasingly scarce.

The incidence of poverty amongst females within households interviewed was found to be somewhat higher than the incidence of poverty amongst males (77 per cent compared with 72 per cent: Table 8.2b). A considerably higher incidence of poverty was found for female-headed households compared with male-headed households (72 per cent compared with 58 per cent: Table 8.2b). These results are interpreted as illustrating that a 'feminisation of poverty' (Government of Zimbabwe, 1997), and it is widely accepted that attempts should be made to actively redress this current gender-imbalance. In terms of land reform, this could include targeting female-headed households for resettlement along with providing training in the necessary skills-base required for successful resettlement activities where this is necessary; and addressing many of the traditional legal constraints to female inheritance, resource ownership, and other factors such as gender-biased constraints to accessing credit (refer to Chapters 14, 15 and 16 for further discussion of these issues).

The Incidence of Poverty by Land-Use Sector

Table 8.3a compares the incidence of poverty in the three main contrasting rural land-use sectors, and the incidence of poverty for the urban land-use sector is repeated here for comparison.

Table 8.3a Poverty incidence for land-use sectors - results of the 1995 PAS Survey

	LSCF	*SSCF & RAs*	*CAs*	*Urban*
% very poor	52	66	77	28
% poor	23	13	12	22
= % total poor	75	79	89	50
% non-poor	24	21	11	50

Source: Government of Zimbabwe, 1997

The results show that communal areas (CA) have significantly the highest proportion of very poor people (77 per cent), and the lowest proportion of non-poor (11 per cent); small-scale commercial farms (SSCF) and resettlement areas (RA) have the next highest proportion of very poor (66 per cent) and second highest proportion of non-poor (21 per cent); the large-scale commercial farms (LSCF) have the lowest proportion of very poor (52 per cent), and the highest proportion of non-poor (24 per cent). The incidence of poverty in urban areas is far less than the incidence of poverty in all rural land-use sectors.

As reported in DANIDA, 1996, this pattern of poverty incidence with land-use sector has not always been so. Table 8.3b shows comparative data for 1990/91, determined from the Income Consumption and Expenditure Survey Report 1990/91 of the 1992 Census (Government of Zimbabwe, 1992), and as reported in the World's Bank 1996 poverty report (World Bank, 1996).

The 1990/91 data indicates that the percentage of poor and very poor living in resettlement areas was actually higher than in communal areas (41 per cent and 48 per cent for RA compared with 33 per cent and 35 per cent for CA: Table 8.3b).

Collaborative evidence includes Thomas, 1991, reporting that the highest rates of stunting were recorded in resettlement areas (Thomas, 1991). In 1994, UNICEF found that wasting was most prevalent in resettlement areas (UNICEF, 1994). In the resettlement areas where this has been a problem, it has been attributed to insufficient institutional support for the activities of those who have been resettled, which is an admitted shortcoming of some of the earlier phases of resettlement of the mid 1980s. The PASS 1995 survey suggests that the situation in the RAs has improved during the 1990s. The research of Kinsey (refer for example, to Chapter 9 of this text) is one such study reporting examples of households in resettlement areas where appropriate institutional support is provided, that now generate higher incomes than households on communal land.

Table 8.3b Incidence of poverty by land-use sector, 1990/91[a]

	% LSCF	%RA	%CA	% Urban
Poverty[b]	16	41	33	10
Extreme poverty[b]	10	48	35	7
Share of poor[c]	6	5	76	12
Share of very poor[c]	4	6	82	8
Share of population[d]	11	4	51	31

a   Determined from the Income Consumption and Expenditure Survey Report 1990/91 of the 1992 Census, Central Statistics Office, Government of Zimbabwe, as reported in the World's Bank 1996 poverty report.
b   Number of poor households in the category / number of households in category.
c   Number of poor households in the category / total number of poor households.
d   Includes other rural sub-groups that account for 2 per cent of the population.

The results presented in Tables 8.3a and 8.3b illustrate that as long as the necessary support services for making resettlement activities productive exist, where land reform resettles people from communal areas into resettlement areas, their resultant land-use activities are likely to result in them being better off.

The incidence of poverty for LSCF has to be interpreted with caution, in that the figures disguise the large measure of household inequality in this sector given the presence of farm owners and high-salaried managers along side the labourers (DANIDA, 1996). The bias on the results is probably limited by the fact that the number of farm labourers far outweigh the number of owners and managers (illustrated by the employment information provided in Table 6.1 of Chapter 6 in this text). Farm labourers are the lowest paid workers in the Zimbabwean formal sector (Amanor-Wilks *et al*, 1995). However, an aspect of their income poverty (measured in the 1995 PASS survey as reported household income) is to varying extents partly alleviated by other non-monetised receipts by the household that is part of the package of their employment, but not accounted for in the methodology used in this 1995 PASS survey. Notwithstanding, the data appears to illustrate much higher incomes on average for households employed on LSCFs compared with households in RA or on CA.

If resettlement displaces the workforce of currently productive LSCF, which is threatened in the current plan of land reform proposed, then whilst those resettled from CA are likely to become better off (if the necessary support services are subsequently provided), any workers resettled from employment on LSCF, depending upon what alternative employment activities are available for them, could become poorer, and rural and /or urban poverty could worsen. Refer to Table 6.1 of Chapter 6 of this text for further information on numbers employed by rural land-use sector that helps to illustrate this point further. The results presented in Table 8.3a also suggest that expanding employment opportunities for households on LSCFs or in urban areas, which is unlikely in the current economic climate, could be an even more effective means of alleviating poverty than land reform, depending upon the mode of land reform undertaken (refer to Chapter 16 of this text for a range of models of land reform suggested).

Influence of Poverty on Views of Adequacy of Grazing and Cropping Land

Within the rural land-use sectors, information was also obtained on whether households considered they had adequate grazing land (Table 8.4), and adequate cropping land (Tables 8.5).

More households considered they had sufficient grazing and cropping land than insufficient (even amongst the most poor families). This suggests that in people's minds there is little link between adequacy of grazing land or adequacy of cropping land with poverty, and suggests that people associate poverty more closely with other factors.

Table 8.4 Views on adequacy of grazing land for each category of poverty - results of the 1995 PAS Survey

	Sufficient	*Not sufficient*	*Not applicable*[a]	*Not stated*
% Very poor	35	22	35	8
% Poor	25	19	45	11
% Non-poor	15	11	55	19

a   Mostly urban-based households.

Source: Government of Zimbabwe, 1997

Table 8.5 Views on adequacy of cropping land for each category of poverty[a] - results of the 1995 PAS Survey

	Sufficient	*Not sufficient*	*Not stated*
Very poor	52	39	10
% Poor	42	36	22
% Non-poor	29	28	44

a   Results from rural households only.

Source: Government of Zimbabwe, 1997

The Influence of Land Ownership on Poverty

Information was obtained on whether or not the household owned land in rural areas. The breakdown of this information for each level of poverty and for each rural land-use sector is presented in Table 8.6.

The results presented in Table 8.6 show a clear inverse relationship between poverty and land ownership: by far the majority of the very poor own land (82 per cent) whereas the majority of the non-poor do not own land (60 per cent). These results may be somewhat biased by the fact that the 1995 PASS survey is based on 'reported household income', excluding non-monetarised gains to the household which would be greater for those obtaining a living from the land. Useful information to qualify this point, however, is obtained from the survey results of household perceptions of causes of poverty and solutions to poverty that is analysed further in Table 8.7.

Table 8.6 Poverty level with ownership of land in rural areas, and rural land-use sector of household with ownership of land in rural areas - results of the 1995 PAS Survey

	Have land	*No land*	*Not stated*
% very poor	82	17	1
% poor	61	37	2
% non-poor	38	60	2
% LSCF	12	84	4
% SSCF & RA	72	27	1
% CA	68	31	1

Source: Government of Zimbabwe, 1997

Perceptions of Poverty

Further insight of the implications for poverty of the Land Reform Programme in Zimbabwe is gained from information obtained as part of the survey study on households' views on what indicates poverty (Table 8.7), what causes poverty (Tables 8.8a and 8.8b), and solutions to poverty (Tables 8.9a and 8.9b). There is ambiguity in the meaning of words commonly used in the translation of 'poverty' or 'the poor' which can make investigations of people's perceptions of poverty quite difficult (DANIDA, 1996). African words for social conditions such as a lack of wealth are closely bound up with conceptions of causality, and as such are considered to be insults, carrying negative connotations of shame, self-pity, and self-blame, which is different to most western uses of these terms. Surveys investigating peoples' views on social conditions thus have to be careful to avoid this confusion. The approach to avoiding such confusion in the 1995 PASS survey was to ask indirectly about the local word conventionally used in translation (DANIDA, 1996).

**Table 8.7 How households identify poverty - results of the 1995
PAS Survey**

	% National	% Urban	% Rural
Shortage of food	73	69	75
Shortage of clothes	11	11	10
Poor accommodation	4	6	2
No draught animals	4	0	6
Poor health appearance	3	4	3
Cannot send children to school	2	3	1
Begging tendencies	2	3	1
Malnourished children	1	1	0
Shortage of farming land	1	1	1
Sample size	18,790	7,315	11,475

Source: Government of Zimbabwe, 1997

In identifying poverty, a 'shortage of food' was by far the most commonly chosen indicator (73 per cent nationally); next came a 'shortage of clothes' (11 per cent nationally), followed by a number of other factors each identified by only a small per cent of those interviewed, with 'a shortage of farming land' being one of the factors least identified (Table 8.7). These results suggest that household food security should be a clear target for government policy and development assistance, no doubt to include enhancing sustainable agricultural production levels, as well as securing food provision by other means, and also investment in factors that, for example, would improve food distribution and storage throughout the country. These results also illustrate the need for land reform to be undertaken in a way that does not threaten, but instead enhances, the sustainable productivity of the country's rural land resource activities.

'Unemployment and retrenchment' was, by far, the most frequently identified causes of poverty in urban areas (44 per cent). In rural areas 'drought' was most frequently identified (40 per cent). 'Low paid jobs' (12 per cent nationally) and 'high prices' (8 per cent nationally) were the next factors most chosen as causes of poverty. Other possible causes, each chosen by just a small per cent of those interviewed, included 'laziness', 'large families', 'poor parents', and 'a lack of education'. Both 'a shortage of land', and 'poor quality land', were each seen to be a main cause of poverty by only 1 per cent of those interviewed (Table 8.8a).

Table 8.8a   Households' perceptions of main causes of household poverty - results of the 1995 PAS Survey

	% National	*% Urban*	*% Rural*
Unemployed/retrenched	30	44	21
Drought	29	10	40
Low-paid jobs	12	14	11
Prices too high	8	13	5
Laziness	7	5	8
Large families with low wages	5	6	4
Poor parents	3	3	3
Lack of education	2	2	2
Ill health	1	0	0
Poor quality land	1	0	1
Shortage of land	1	0	1
Rural/urban migration	0	0	0
Sample size	*18,790*	*7,314*	*11,475*

Source: Government of Zimbabwe, 1997

These results suggest that maintaining employment opportunities and job creation schemes should be an important focus for government policy and development assistance in urban areas; while drought mitigation measures such as improved water resource management and irrigation development should become a priority in rural areas. In other words, an improved support of agricultural productivity in rural areas was still needed regardless of land reform, and would also be essential to secure the productivity of resettlement activities where land reform takes place.

Table 8.8b Households' perceptions of main causes of poverty by rural land-use sector - results of the 1995 PAS Survey

	% National	*% Urban*	*% Rural*
Unemployed/retrenched	25	22	20
Drought	12	39	52
Low-paid jobs	35	7	2
Prices too high	9	7	3
Laziness	4	10	10
Large families with low wages	8	4	3
Poor parents	2	3	3
Lack of education	1	2	2
Ill health	1	1	1
Poor quality land	0	1	1
Shortage of land	2	2	1
Rural/urban migration	0	0	0
Sample size	*2,952*	*1,051*	*7,472*

Source: Government of Zimbabwe, 1997

When analysed by rural land-use sector (Table 8.8b), it is interesting that the above patterns of responses remain similar: 'drought' being by far the most often perceived main cause of poverty in both communal areas and resettlement areas, but 'unemployment / retrenchment', and 'low paid jobs' being most often perceived as the biggest cause of poverty by those in large-scale commercial farms. Again, 'poor quality land' and 'a shortage of land' were chosen as a main cause by only 0 per cent to 2 per cent of those interviewed in all three categories (1 per cent and 1 per cent by those on communal land, 0 per cent and 2 per cent by those on LSCFs, and 1 per cent and 2 per cent by those on resettlement land). These findings provide further support to the inferences made above.

The results of the survey exploring households' views on solutions to poverty are presented in Table 8.9a (the national response, including a breakdown for rural compared with urban areas) and Table 8.9b (by rural land-use sector).

Table 8.9a   Households' perceived solutions to household poverty - results of the 1995 PAS Survey

	% National	*% Urban*	*% Rural*
Create employment	24	35	17
Salary/wage increase	15	17	13
Affordable agri. loans	13	3	19
Irrigation water	11	2	16
Project support fund	10	12	8
More cooperatives	4	5	3
Self-reliance	4	5	3
Encourage income generation	3	4	3
Provision of accommodation	2	4	1
Provision of education	2	3	2
Improve infrastructure	2	1	3
Provision of land	2	1	3
Better health facilities	2	2	1
Establish training facilities	1	1	0
Sample size	*18,790*	*7,314*	*11,475*

Source: Government of Zimbabwe, 1997

The results for the nation as a whole, and for urban versus rural areas, reveals that the 'creation of employment', and 'salary/wage increases' were by far the most frequently perceived solution to household poverty both nationally (24 per cent and 15 per cent), and in urban areas (35 per cent and 17 per cent). These two solutions were also put forward by households in rural areas. A slightly more popular choice for rural areas was 'affordable agricultural loans' (19 per cent for rural areas and 13 per cent nationally:

Table 8.9a). In urban areas the third most frequently chosen solution to poverty was 'project support funds' (12 per cent for urban areas and 10 per cent nationally). 'Provision of land' was amongst the least chosen options for solving household poverty both nationally (2 per cent), in urban areas (1 per cent), and in rural areas (3 per cent).

Table 8.9b   Households' perceived solutions to poverty by rural land-use sector - results of the 1995 PAS Survey

	% National	% Urban	% Rural
Create employment	20	17	16
Salary/wage increase	44	7	2
Affordable agricultural loans	4	21	25
Irrigation water	2	18	22
Project support fund	5	10	9
More cooperatives	2	2	4
Self-reliance	3	3	3
Encourage income generation	2	2	4
Provision of accommodation	3	3	0
Provision of education	2	3	2
Improve infrastructure	1	1	3
Provision of land	5	3	2
Better health facilities	1	1	1
Establish training facilities	0	0	0
Sample size	*2,952*	*1,051*	*7,472*

Source: Government of Zimbabwe, 1997

By rural land-use sector (Table 8.9b), those in communal areas and resettlement areas chose 'affordable agricultural loans' (25 per cent and 21 per cent) and 'irrigation water' (22 per cent and 18 per cent) as the most popularly perceived solutions to poverty. 'Employment creation' and 'higher wages' remained preferred solutions to poverty for households on large-scale commercial farms (20 per cent and 44 per cent). For all rural land-use sectors, the 'provision of land' was amongst the less chosen solution to poverty (chosen by only 2 per cent of households in communal areas, by 5 per cent of households on large-scale commercial farms, and 3 per cent of households in resettlement areas).

In terms of poverty alleviation, these results re-emphasize the need for securing jobs and for job creation. In terms of land reform, the need for a focus on the provision of infrastructural support for resettlement activities and other non-resettled agricultural activities (such as schemes for providing affordable agricultural loans, and irrigation development) are shown to be more popular solutions to poverty than land reform itself, and the dangers of resettled families and other agriculturalists not being offered such support are illustrated.

Conclusions

The findings of the 1995 Poverty Assessment Study Survey presented indicate that unacceptably high levels of poverty are occurring throughout Zimbabwe. The study is invaluable for the detail it gives not only of the demographic, geographic and sectoral distribution of poverty in Zimbabwe, but also for the insight it gives into the interrelatedness of poverty, and perceptions of poverty by the Zimbabwe people suffering from poverty today. The findings can be used as an invaluable guide to identifying what the Zimbabwe people perceive as most needed focuses for international donor assistance targeting poverty alleviation, and for government policy.

The findings of the survey illustrate a very clear belief amongst the 19,173 households interviewed that the more immediate causes of poverty lie more in 'unemployment', 'low wages' and 'drought' than either 'poor land quality' or 'a shortage of land'. Furthermore the households interviewed perceive the more immediate solutions to poverty as being the 'creation of employment' and 'higher wages' in urban areas, and 'affordable agricultural loans' and the 'provision of irrigation' in rural areas, more than the 'provision of land'. This does not imply that land redistribution cannot also help alleviate poverty in Zimbabwe, and important empirical evidence of how land reform undertaken in the 1980s has significantly improved rural welfare is presented in Chapter 9 of this text. The findings of the study reported here help illustrate what model of land reform is likely to be most successful in this regard and why. They also suggest, however, that it would be neither popular by the people of

Zimbabwe that such action would be intending to assist, nor successful, if land resettlement alone was the only action that was being relied upon to alleviate and reduce poverty in Zimbabwe.

The findings also illustrate throughout the critical role of an adequate provision of support services for all agricultural activities (for example, the need for affordable agricultural loans to help resettlement activities get underway, and to assist in improving productivity levels on communal land amongst those left out of the resettlement process as well; and the provision of irrigation in rural areas to help overcome the effects of drought and to improve yields, again both for subsequent resettlement activities, and for those left in communal areas), if vital goals (for example, those of productivity and equity) of the Land Reform Programme are to be successfully met at all.

These findings should be considered in assessments of what government policy and development assistance should target in the future, if productivity of the precious land resource base of the country is to be maintained, enhanced, and sustained.

Note

1   The basket of basic food being those that satisfy the calorie and protein requirements for both rural and urban areas (2100 calories on average per person per day recommended by the FAO and WHO), based on an analysis of the nutritional content of the standard consumption of a reference population group; and using average consumer prices for each of the items in the basket (refer to Government of Zimbabwe, 1997, for further details).

References

Amanor-Wilks, D. *et al.* (1995), *In Search of Hope for Zimbabwe's Farm Workers*, Dateline Southern African & Panos Institute Press, London.

DANIDA (1996), *Evaluation of Poverty Reduction in Danish Development Assistance, Country Study: Zimbabwe*, Main Report: vol.1, Ministry of Foreign Affairs, Government of Denmark.

Government of Zimbabwe (1992), *Population Census 1992: National Report*, Central Statistics Office, Government Printers, Harare.

Government of Zimbabwe (1995), *Zimbabwe's Agricultural Policy Framework 1995 - 2020*, Ministry of Agriculture, Government Printers, Harare.

Government of Zimbabwe (1997*), Poverty Assessment Study Survey Main Report*, Ministry of Public Service, Labour and Social Welfare, February 1997, Government Printer, Harare.

Her Majesty's Government (1997), *Eliminating World Poverty: A Challenge for the 21st Century*, UK Government White Paper on International Development. Presented to Parliament by the Secretary of State for International Development, by Command of Her Majesty, November 1997, HMSO, London.

Moyo, S. (1994), *Economic Nationalism and Land Reform in Zimbabwe*, Occasional Paper Series No. 7, SAPES Trust, SAPES Books, Harare.

OECD/DAC (1996), *Shaping The 21st Century: The Contribution of Development Cooperation*, OECD, Paris.

OXFAM (1996), *The OXFAM Poverty Report*, OXFAM Publications, London.

Thomas, D. (1991), *Household Resources and Child Health in Zimbabwe*, University College London / Yale University, New Haven Press, London.

UNDP (1996*), Human Development Report*, Oxford University Press, New York.

UNICEF (1994), *Children and Women in Zimbabwe: A Situation Analysis – Update,* United Nations Children Fund, Harare.

World Bank (1996), *Understanding poverty in Zimbabwe: Changes in the 1990s and Directions for the Future*, Harare and Washington, June 1996.

9 The Implications of Land Reform for Rural Welfare

DR B. H. KINSEY

Introduction

Zimbabwe's Land Reform Programme, introduced in 1980 just months after the liberation war ceased, has been widely criticised. Most of these criticisms are paradoxical in several respects, however. First, a considerable body of received wisdom advises that the benefits of programmes which involve large-scale human resettlement are unlikely to become apparent in less than a generation (Colson, 1971; Nelson, 1973; Scudder, 1973, 1975; Kinsey and Binswanger, 1996). Moreover, full economic maturity of even the earliest phase of Zimbabwe's own programme will not be attained until the year 2000. The sweeping judgements on the programme which began to appear within a few years of its inception have thus all been premature.

Further, the resettlement programme has been more complex and diverse than most. It involves approaches that emphasise uniform family-based holdings (known as Model A), collective cooperatives (Model B), and links between satellite producers and centralised commercial crop and livestock production and processing (Model C - a rarity). It also includes efforts - continuously modified - to devise an approach to resettlement which suits the needs of the semi-arid parts of the county, where livestock-keeping appears to offer more promise than cropping (originally Model D, now the 3-tier model - also a rarity) .

Most of Zimbabwe's Land Reform Programme has been implemented through Model A. Of the more than 71,000 families resettled to late 1996, some 93 percent have been resettled on Model A schemes (Government of Zimbabwe, 1996a). It is paradoxical that the Model A schemes, which have consumed the lion's share of resources and which have the greatest aggregate potential for alleviating poverty, have received so little systematic attention. A second paradox is that no critique has addressed fully the programme's original set of objectives. It is worth remembering that three of the seven main objectives of the programme are poverty alleviation, welfare enhancement and national stability (Government of

Zimbabwe, 1980). This chapter focuses on Model A resettlement. The analysis takes the original - largely political - objectives of the programme, which placed great emphasis on welfare and poverty reduction, and assesses the extent to which these have been met at the level of the household. The major emphasis is on comparing indicators of 'growth' among resettled households with a 'control group' and assessing changes in income over time.

The Scale of Redistribution

Zimbabwe's programme dwarfs all prior voluntary resettlement programmes in sub-Saharan Africa. Some 3.5 million hectares of land, 84 percent of it acquired from the large-scale commercial farming subsector, had been committed to the programme between 1980 and late 1996 (Government of Zimbabwe, 1996b). In terms of accomplishments compared to target numbers and deadlines, however, Zimbabwe's programme lags far behind its own implementation schedule.

The programme can also be criticised for the dilution over time of its poverty-alleviation goal. Changes in selection criteria in 1990 to emphasise farming experience and possession of agricultural capital led one observer to conclude, 'There has been...an abandonment of the earlier commitment to redistributive justice' (Dashwood, 1996). Despite clear evidence that what has been called 'the myth of unproductive peasants' (Weiner *et al*, 1991) is merely that - a myth, the state also appears happy to preserve a dualistic structure through transferring title in large holdings from white to black owners.

Target-setting has haunted the programme. The initial target, set in 1980, was 18,000 families to be resettled on 1.1 million hectares of land at a cost of Z$60 million. This figure was derived not from an assessment of need, but rather from calculations of what could be accomplished under the joint British-Zimbabwean programme launched after independence. This initial modest target proved attainable and, by 1987, 40,000 households had been resettled on two million hectares (Cusworth and Walker, 1988).

Within two years of independence, a revised target of 162,000 households (on nine million hectares at a cost of Z$500 million) was published in the Transitional National Development Plan (Government of Zimbabwe, 1982). While this figure was acknowledged to be no more than a guess (Government of Zimbabwe, 1996a), it proved to be wildly optimistic in terms of what would be accomplished. Commenting on the enduring failure to meet targets despite a rhetorical commitment to land redistribution, one observer notes (Goebel, 1998):

> While government primarily blames lack of funds for land purchase and infrastructural development, other factors stalling the establishment of sufficient numbers of schemes to meet demand include the centrality of commercial agriculture to the Zimbabwean economy, pressure from multilateral institutions to protect commercial farming, government corruption through which acquired land is given to ministers and other government friends, and flagging commitment on the part of the new black élite housed in government to the welfare of the poor masses.

Fifteen years after it first became operational, Zimbabwe's resettlement programme 'had never been evaluated to determine the strengths and weaknesses as well as the need for its redesign' (Government of Zimbabwe, 1995). Despite the absence of a comprehensive review, major developments took place throughout this period which affected significantly the policies and procedures guiding the programme. Examples include the first and second 5-year national development plans in 1986 and 1991, the adoption of a new land policy in 1990, enactment of a new land acquisition act in 1992 and its revision in 1996, the introduction of structural adjustment in 1991, the hearings and report of the Land Tenure Commission (LTC), and - most recently - major revisions to land policy (Government of Zimbabwe, 1996b).

The substantial redesign of the programme also occurred in the face of inadequate evidence on what was happening at the level of the resettled household. Survey programmes were never given sufficient resources to operate on more than an *ad hoc* basis, and most so-called evaluations were in fact merely reviews of the progress of implementation. It is not surprising, therefore, that the redesign failed to address many of the problems of the resettlement programme.

Indicators linking Resettlement to Welfare Improvements

The resettlement programme was implemented very rapidly during the early 1980s but has slowed dramatically since that time. The reasons why the pace of the programme has changed are complex and beyond the scope of this analysis. Among them, however, are a barrage of criticisms and negative evaluations - from both within and outside government - that the programme has failed to have a positive impact on agricultural productivity and rural incomes.

Zimbabwe's resettlement programme was planned in the atmosphere of the *growth-with-equity* theme prominent in development thinking in the late 1970s and early 1980s, and the declared objectives of the early programme were couched in such rhetoric. It seems appropriate therefore to choose as measures of the welfare gains from resettlement indicators which quantify both dimensions of economic performance as well as those which tell us something about the distribution of growth.

There are a number of measures of inequality which the analyst of distributional issues might employ. Perhaps the most commonly used is the *Gini coefficient*. The Gini coefficient ranges from zero to one. Value of zero reflects complete equality - all individuals benefit equally. At the other extreme, a value of one reflects the case where a single individual receives all benefits and no one else receives any. Thus, generally speaking, the larger the Gini coefficient, the greater the inequality.

Resettled households are no longer typical of rural households in Zimbabwe. Although families were initially meant to be selected for resettlement because they were landless, displaced by war or relatively poor, they have been given access to superior resources and have used it to build up a base of assets, particularly livestock. This accumulation of assets has been detailed elsewhere (Kinsey *et al*, 1998), but the outcome is that 90 per cent of households now own cattle, with an average holding of some 10 animals. Moreover, the programme has provided settlers with a wide range of facilities, such as water supplies, diptanks, clinics, schools, improved toilets, housing loans, roads and marketing depots. Critical services - such as clearing and ploughing part of each settler's arable land - have also been provided.

The Data

The data used for this analysis come from two sources. During 1982-84, a panel was constructed from households resettled in the first two years of the programme - 1980-82 - in Zimbabwe's three agriculturally most important agroclimatic zones (known as Natural Regions - NRs). Households in the panel reside in NRII, NRIII and NRIV, respectively areas of moderately high, moderate and low or restricted agricultural potential. One scheme was selected from each zone. The resulting panel, which has been surveyed annually since 1983, comprises some 400 families residing in 20 villages. All assessments of change over time draw upon the panel study. In these cases, the data used represent the initial year (1983/84) and one or more 'terminal years'.

Second is a set of data especially assembled in 1997 to provide a contrast to the resettlement experience. For each resettlement area (RA), the two communal area (CA) villages which had supplied the largest number of panel households were selected. In each selected village, 25 households were interviewed. Thus, for 1997, the complete data set comprises 397 resettled households in the panel and a 'matched set' of 150 households from the CAs.

Table 9.1 Characteristics of households, three RAs and six CAs, 1997

Characteristic	RAs (n=397)	CAs (n=150)
Household size		
Mean no. of residents	10.92	6.63
Range: min.-max. no. of residents	2-41	1-18
Sex composition		
Male	51.6	48
Female	48.4	52
Age profile		
Mean age	21.6	23.2
Mean activities of household heads (%)		
Farming	95.5	80
Unable to work	2	1.3
Other	2.5	18.7
Human capital		
Mean no. of years of education per capita	4.8	4.4
Farming		
Hectares planted (1995/96)	3.5	1.7
Total revenue from crop sales (Z$)	10,949	1,614
Revenue per hectare planted (Z$)	2,778	852

Source: Kinsey 1998

Characteristics of the households and heads of households for the three RAs and six CAs are summarised in Table 9.1. The most important differences between RAs and CAs are the much larger family sizes in RAs, the smaller areas planted in CAs and their apparent lower productivity, and the proportion of household heads who report farming as their main activity.

Comparative Evidence on Household-level Welfare

This section compares a set of indicators of welfare in terms of both absolute levels and distributional inequality for households in both RAs and CAs.

Farm Incomes

Since the planning for resettlement indicated that all quantified benefits from the programme would arise from farming activities, the comparison here is limited to measures of income and wealth derived from crops and livestock.

Several comparisons are set out in Table 9.2. The first values the entire crop output of the household from the relatively good 1996 harvest at the median price received by those who sold the crop. Thus, production for own-consumption is valued the same as production for sale. On this basis, the crop output of the average resettled family is worth over four and a half times that of the average CA household.

A similar comparison has been made for that proportion of crop output which was actually marketed. In this case, however, the comparison was made in terms of actual revenue received. Here the difference is even more striking. Resettled farmers earned from crop sales 6.8 times what CA farmers earned. The differences in proportions of output marketed are also striking. Resettled households sold 78 per cent of the value they produced, whereas CA households marketed only 53 per cent of the value of their output.

It is not uncommon to find that as mean incomes rise so also does the variability in incomes, and income distribution may worsen as well. Table 9.2 also contains an indicator of variability of income. The coefficient of variation (the standard deviation divided by the mean and expressed in percentage terms) shows that while incomes are very variable, they are markedly more variable in CAs than in RAs.

Finally, the *Gini coefficient* provides a measure of distributive inequality for crop income for the two groups of farmers. With both measures, the markedly higher values of crops in resettlement areas are accompanied by lower Gini coefficients, and the difference between the two areas in marketed output is particularly striking. Similar comparisons are reported in Table 9.2 for livestock. Here the value of the average livestock holding in RAs is roughly double that in CAs.

Table 9.2 Distributive inequality measures for values of crop production, cropping season 1995/6 and livestock 1997, RAs and CAs

	RAs (*n=397*)		CAs (*n=150*)[a]	
Crops	*Sale value*[b]	*Revenue from marketed output*	*Sales value*[b]	*Revenue from marketed output*
Mean (Z$)	14,027	10,949	3,045	1,614
Coefficient of variation	101	109	121	176
Gini	0.49	0.58	0.52	0.74
Livestock	*Value of holdings*[c]	*Revenue from livestock*[d]	*Value of holdings*[c]	*Revenue from livestock*[d]
Mean (Z$)	20,508	188	10,125	56
Coefficient of variation	104	326	110	264
Gini	0.47	0.83	0.54	0.88

a   The number of observations for communal area livestock is 149 due to the exclusion of one large-scale commercial poultry producer.
b   The value of total crop output valued at the median per-unit price for all households which actually sold the crop.
c   The household's aggregated holdings of large and small stock with each type of animal valued by the farmer at its estimated market value at the time of interview.
d   The sales revenue from livestock, livestock products and services.

Source: 1997 Panel survey round

Mean values of sales of livestock and livestock products (eggs, milk, skins, etc.) and services (ploughing and transport) are very low in comparison to crop sales and are extremely variable, although these outcomes are likely to be sensitive to the season in which the data were recorded. The average resettled farmer earns some 3.4 times more from livestock-related revenue than his CA counterpart.

On the basis of the Gini coefficients, the same pattern is revealed for livestock as for crops. That is, higher values and revenues are associated with more equal distributions of income.

Income from Off-Farm Sources

Although the planners who designed the resettlement programme assumed that households would earn their entire income from farming activities, they ignored the importance of transfers and remittances, common throughout southern Africa. Such transfers commonly take place in both cash and kind, but only the former are considered here.

Table 9.3 reports the distributive inequality measures for both cash remittances and other cash transfers. Recalling that RA plot-holders were for years prohibited from holding non-farm jobs, it would be expected that remittances to RA households would be less than among CA households, where labour migration is far more common. This is the pattern that emerges.

Total cash transfers to CA households are some 60 per cent higher than those to RA households. Because CA households are about a third smaller than their RA counterparts, it follows that transfers contribute relatively more to total household welfare in CAs than in RAs. The greater importance of transfers to CA households is further reinforced by the fact that, despite the higher absolute level, transfers to CA households are more uniform (as indicated by the smaller coefficient of variation) and more equitably distributed (the Gini coefficient). Transfers to both sets of households are nevertheless extremely variable and very inequitably distributed.

Expenditure for Non-Food Items and for Food

The preceding discussion has addressed the major sources of household incomes. Consumption however is a better indicator of welfare than income because consumption tends to vary little from month to month whereas income usually has pronounced seasonal influences. Total consumption may be a more reliable indicator of welfare than food consumption because poorer households tend to spend a large proportion of their income on food than those that are better off.

Table 9.3 Distributive inequality measures for remittances and cash transfers, RAs and CAs, 1996/7*

	RAs (*n=397*)	CAs (*n=150*)
Mean (Z$)	581	932
Coefficient of variation	337	216
Gini	0.86	0.79

* Includes cash payments of bridewealth; gifts, assistance and aid; inheritance; pensions and remittances in cash. Excludes remittances and transfers in kind.

Source: Kinsey 1998.

Non-food expenditure was recorded for the 12-month period preceding the 1997 survey. This expenditure is used here as a measure of consumption. The results are presented in Table 9.4 in two ways: all household expenditure and all expenditure excluding education. In both cases, reported expenditure for RAs is some 1.6 times that for CAs. Non-food expenditure in RAs is also much less variable than in CAs and is more equitably distributed.

By reporting expenditure on a per-household basis, Table 9.4 conceals as much as it reveals. If the figures for total expenditure in Table 9.4 are divided by the mean household sizes in Table 9.1, the result is a mean per-capita expenditure of Z$644 in CAs compared to Z$638 in RAs. The result stands, however, that consumption in RAs is more equitably distributed than in CAs.

Recalling that a food-based measure may understate consumption, and hence welfare, we turn now to just such a measure. In each year, food expenditure has been recorded in detail for the month preceding the survey. The survey normally coincides with the period when food stores can be expected to be near their lowest.

Table 9.4 reports the results for food expenditure following the relatively good 1996 harvest. The results fit expectations. Land-short CA households spend nearly four times as much as RA households for grain, although the expenditure levels for both are low. The very large coefficients of variation and Ginis for both indicate that only a small proportion of households in each tenure category bought grain in early 1997. In this case, the Gini and coefficient of variation for RAs should be interpreted as indicating the need to purchase grain in RAs is less widespread than in CAs.

Purchases of all foodstuffs are remarkably more uniform on a per-capita basis despite the differences at household level in Table 9.4. Although the mean expenditure per RA household is 1.4 times that of the CA household, on a per-capita basis the average RA household spent Z$36.91 on food while the average CA household spent Z$ 44.06.

Taken together, the various indicators of income discussed above provide convincing evidence that the land reform beneficiaries of the early 1980s have, by the late 1990s, used the land provided them to good advantage. Advantage here translates to a comparatively strong economic performance from agriculture and livestock and more favourable distributions of income and consumption. RAs have realised their potential to operate as viable small-scale commercial farming areas while the CAs from which the settlers came, in contrast, display a continuing reliance on non-agricultural sources of income.

Beneath the impressive economic performance which older RAs are beginning to display are, however, some indicators that income is not reliable as the sole indicator of welfare. There is, for example, no convincing evidence that land reform has improved child welfare as assessed by nutritional indicators. Quite the contrary. A paradoxical outcome is that child nutrition is inversely correlated with household income; children in low-income households in marginal agro-ecological zones are systematically better nourished. Additional land has led to greater cash-cropping, more expenditure on inputs and higher mean incomes, that could be used to reduce chronic undernutrition in RAs. It seems clear, however, that for many families not only is chronic undernutrition unaffected, but also they remain unprotected against periodic food crises.

Longitudinal Evidence on Household-level outcomes

The preceding discussion has focused on the analysis of household-level differences between beneficiaries of resettlement and non-beneficiaries residing in CAs from which many beneficiaries originated. Since the comparisons are based on results from a single, reasonably good season, it is of greater interest to explore what has happened to RA households since they were resettled.

Table 9.4 Distributive inequality measures for non-food expenditure and for monthly food purchases, RAs and CAs, 1996/97*

	RAs (n=397)		CAs (n=150)	
Non-food expend.	*Expend. excl. education*	*All expend.*	*Expend. excl. Education*	*All expend.*
Mean (Z$)	5,984	6,974	3,705	4,271
Coefficient of variation	88	84	124	113
Gini	0.39	0.38	0.48	0.45
Food purchases	*Purchases of grain*	*All foodstuffs*	*Purchases of grain*	*All foodstuffs*
Mean (Z$)	4.64	403.07	17.24	292.14
Coefficient of variation	540	91	400	102
Gini	0.97	0.40	0.91	0.45

*   Expenditure excludes acquisition of agricultural and household assets. Reported food purchases are for a one-month period during January-April 1997. Excludes food acquired through barter and food purchased and consumed away from home.

Source: Kinsey 1998

Table 9.5 summarises changes in real income for the period covered by the panel study, taking the baseline season - 1982/83 - the initial year and three different 'terminal' years. The results for each terminal year reflect very different cropping seasons and illustrate clearly the risks involved in employing single cross-sectional surveys to assess outcomes of programmes which are long-term in nature. Depending upon the terminal year chosen, one might conclude that mean incomes had increased by 321 per cent, 132 per cent or 465 per cent. Since the baseline data were collected in a bad season and 75 per cent of the years since resettlement began have experienced below-normal rainfall, one might want to restrict comparisons to like-with-like and choose the lowest figure, that for 1994/95. The conclusion would then be that average real incomes have more than doubled over the period and, moreover, have become less variable.

Table 9.5 Changes in real income per household, baseline and three most recent growing seasons

Growing season	No. of households	Mean income[a] (in 1990 Z$)	Coeff. of variation	Min - Max
1982/3[b]	356	765	151	0 - 8,428
1993/4[c]	397	3,219	108	120 - 26,083
1994/5[b]	393	1,778	116	0 - 20,829
1995/6[d]	397	4,324	97	17 - 35,553

a Income is the sum of the market value of crops harvested, revenue from sales of livestock and livestock products and services, remittances and income from off-farm or non-agricultural employment. Excluded are other transfers, aid and drought relief.

b Season was badly affected by drought; rainfall was less than two-thirds of the long-term mean.

c Season was drought-affected; rainfall was some 80 percent of the long-term mean.

d Season was normal; rainfall was within one percent of the long-term mean.

Source: Panel survey, various years

However, a large part of the story of what has happened to incomes is not revealed by the figures in Table 9.5. Underlying the large annual variability in incomes has been a very substantial amount of investment and accumulation of wealth, principally in the form of cattle. Between 1983 and 1995, households in the panel increased the real value of their holdings of cattle by some 200 per cent, despite periodic losses to drought, and sales to smooth consumption (Kinsey *et al*, 1998).

The impact of higher income levels can be assessed by looking at the way past expenditure patterns have improved the welfare of households. In 1983/84, an index of material well-being was constructed for each household. This index assigned a score to each of seven items: housing, domestic water supply, sanitation facilities, cooking fuel, source of lighting, home furnishings and farm and transportation equipment. The identical exercise was repeated almost a decade later, in 1992/93.

The index of material well-being increased by 52 per cent over the period and displayed less variability at the end than at the beginning. Moreover, no household was at the minimum value of the index in 1992/93. While all households improved their positions over time, the average index increased by 61 per cent for the bottom 30 per cent while it rose by only 47 per cent for the top 30 per cent, indicating relative gains by those less well-off. Distributive inequality has declined over time along a scale indicating material well-being.

Discussion and Conclusions

The main purpose of this paper has been to show that Zimbabwe's resettlement programme has, with time, resulted in improved rural welfare in the form of both (i) more equally distributed incomes and (ii) higher incomes. Resettled households crop twice the amount of land and earn more than three times the unit revenues of CA families. Values of livestock, crop production and food and non-food expenditure are all higher and more equitably distributed in RAs than in the neighbouring CAs. Further, the average RA household relies far less on cash remittances and spends less on staple cereals than its CA counterpart.

Over time, resettled households have increased their incomes, and also reduced income variability, while at the same time accumulating considerable wealth in the form of cattle. Although not documented here, they have also invested heavily in housing and other domestic assets. In addition, the measure used for material well-being shows, over the decade covered, both a dramatic increase and an improvement in the distribution.

In broad terms, then, the indicators used agree in showing that the early beneficiaries of Zimbabwe's resettlement programme are substantially better-off than their neighbours who were not resettled. The early beneficiaries are also much better-off than they were shortly after they were resettled. There are other, less-easily-quantified indicators of welfare improvements also. For example, police, social workers and others frequently remark that there is less alcoholism in RAs than other rural areas, less domestic violence and fewer suicides.

It is disturbing, however, that an important indicator of well-being - child nutrition - has not shown improvements in parallel with other indicators. It is also worrying that the shift toward high-value cash crops appears to be associated both with an increase in polygymy and an accompanying increase in divorce. Finally, there is evidence that RAs mirror a wider demographic change, perhaps worsened by structural adjustment, towards a delayed age of marriage for young men. Because older male children cannot easily obtain their own plots in RAs, RA households continue to grow and benefits will with time become more and more diluted.

As noted earlier, conclusions based on evaluations of land reform over short horizons are likely to miss the beneficial impacts that emerge only as communities organise themselves, engaging fully within their communities and with the outside world of support services and economic agents. Predicting the time horizon for such organisation and engagement is a hazardous undertaking. With such uncertainty regarding timing, especially in a context of great seasonal variability, conclusions based only on a single survey are likely to mis-state seriously the outcomes arising from land reform. This analysis has combined both panel and comparative data in an attempt to counterbalance the risk of the one-shot conclusion.

The chapter provides evidence that land reform has significantly enhanced welfare for beneficiaries from the earliest phases of resettlement. Yet there appears to be remarkably little recognition at the policy level that, through land reform, Zimbabwe has the potential to channel the energies of tens of thousands of small-scale farming households into significant contributions to national development. Such a broad-based approach would do far more in the long-term to increase agricultural output and raise incomes than the current approach, which is so heavily rooted in narrow political patronage.

Moreover, as the analysis shows, broad-based land reform leads to declining levels of inequality, a fact likely to reinforce political stability. When the incomes of a political or economic élite are high or increasing rapidly while the incomes of rural non-élites stagnate, there is a risk that large segments of the population will become politically alienated. Declining inequality implies, in contrast, that non-élites are sharing in the benefits of economic growth. When the benefits of growth are shared, there is less risk of political alienation of large parts of the population. Zimbabwe currently faces a crisis of political consensus and threatens to deal with it through a massive confiscation of land and a rapid 'completion' of the resettlement programme. If the threats are carried out, there is likely to be subsequently an even greater economic crisis.

It is appropriate to conclude, not with certainty but with five questions, in the hope that these will be addressed as government seeks to find a way forward from the awkward position it has put itself in:

- Why has government apparently abandoned the poverty focus of the original resettlement programme just when it is reasonable to expect the full effects of the early efforts to begin to manifest themselves?
- How can government argue that it has the resources to support adequately a vastly enlarged resettlement programme when it has never fully supported the programme begun in the early 1980s?
- Why has government not invested the resources to ensure that it learns what works well in the existing programme and what does not work at all? How are we to capture the lessons of experience and correct past mistakes?
- What is the urgency, in Robert Mugabe's terms, to 'finish' land reform? Land reform is a dynamic, evolving process that is finished only when an economy collapses completely. Land reform is better sold politically as a continuing part of a country's development programme than as a finite exercise to be finished within a prescribed time horizon. Someone should tell the President.
- Finally, government, although it may not know it - or care to believe it - has a success story on its hands. Why not let this story be told by those who have experienced genuine welfare benefits from land reform?

Zimbabwe's experience with land reform and resettlement also teaches that it is dangerous to attach planning targets too tightly to political platforms because so many uncontrollable factors - such as drought - can affect the achievement of targets. Moreover, land reform involves very complex social and economic dynamics, which vary both with time and place; 'blanket' approaches are seldom likely to work on any large scale. It is important therefore to attempt to tailor resettlement programmes to specific local circumstances and to allow progressive modification. Models that would allow such flexibility are outlined in Chapter 16 of this text.

References

Colson, E. (1971), *The Social Consequences of Resettlement*, Manchester University Press, Manchester.

Cusworth, J. and Walker, J. (1988), 'Land resettlement in Zimbabwe', *ODA Evaluation Report EV 434*, Overseas Development Administration, London.

Dashwood, H.S. (1996), 'The relevance of class to the evolution of Zimbabwe's development strategy, 1980-1991', *Journal of Southern African Studies* 22, 1, 27-48.

Goebel, A. (1998), 'Process, participation and power: Notes from "participatory" research in a Zimbabwean resettlement area', *Development and Change* 29, 2, 277-305.

Government of Zimbabwe (1980), *Resettlement policies and procedures*, Ministry of Lands, Resettlement and Rural Development, Harare.

Government of Zimbabwe (1982), *Transitional National Development Plan, 1982/83 - 1984/85*, Government Printer, Harare.

Government of Zimbabwe (1995), *Review of the Land Resettlement Programme, 1980-1995*, Ministry of Local Government, Rural and Urban Development, District Development Fund, Harare.

Government of Zimbabwe (1996a), *Memorandum to Cabinet on Resettlement*, Ministry of Local Government, Rural and Urban Development, Harare.

Government of Zimbabwe (1996b), *Policy Paper on Land Redistribution and Resettlement in Zimbabwe*, Ministry of Local Government, Rural and Urban Development, Harare.

Kinsey, B.H. (1998), *Land reform and the implications for rural welfare: The way forward for Zimbabwe's resettlement programme*, paper presented at the Conference on Land Reform in Zimbabwe, 11 March 1998, School of Oriental and African Studies, University of London.

Kinsey, B.H. and Binswanger, H.P. (1996), 'Characteristics and performance of agricultural resettlement programmes', in J. van Zyl, J. Kirsten and H. Binswanger (eds.), *Policies, Markets and Mechanisms for Land Reform in South Africa*. Oxford University Press, Cape Town.

Kinsey, B.H., Burger, K. and Gunning, J.W. (1998), 'Coping with drought in Zimbabwe: Survey evidence on responses of rural households to risk', *World Development* 26, 1.

Nelson, M. (1973), *The Development of Tropical Lands: Policy Issues in Latin America*, Johns Hopkins University Press, Baltimore.

Scudder, T. (1973), 'The human ecology of big projects: River basin development and resettlement', in B. Siegel (ed.), *Annual Review of Anthropology*, Annual Reviews Inc., Palo Alto.

Scudder, T. (1975), 'Resettlement', in N. F. Stanley and M. P. Alpers (eds.), *Man-made Lakes and Human Health*, Academic Press, London.

Weiner, D., Moyo, S., Munslow, B. and O'Keefe, P. (1991), 'Land use and agricultural productivity in Zimbabwe', in N. D. Mutizwa-Mangiza and A. H. J. Helmsing (eds) *Rural Development and Planning in Zimbabwe*, Avebury/Gower Publishing Company Ltd, Aldershot.

10 Zimbabwean People's Perceptions of the Land Resettlement Programme: the Case of Rural-Urban Migrants

DR DEBBY POTTS

Introduction

By the mid 1990s a very significant body of research on the nature of Zimbabwe's land reform programme had been published, and much more was available in the grey literature (e.g. government and agency documents) or was completed but unpublished (e.g. much of Kinsey's longitudinal work comparing resettlement areas to communal areas from the early 1980s, see Chapter 9 of this text). However, relatively little attention had been paid in the published academic and official literalture, to what the people of Zimbabwe felt about the land resettlement programme.[1] Early work by Kinsey (1984) found that settlers who had moved on to schemes in 1980 and 1981 regarded them as a 'mixed blessing', and most felt their families were worse off than those not on schemes. On the other hand, only 5 per cent wanted to return to their home areas, and, as the surveys were conducted after the first post-independence drought, this very clearly affected people's attitudes and the problems they were then facing. Kinsey's subsequent work (Chapter 9 of this text; 1999) shows quite clearly that by the 1990s, settlers are on the whole deriving significant economic advantages from their new land, and are distinctly better off economically than their communal area counterparts. Settlers' perceptions are also covered to some extent in research by Berry (1991), Jacobs (1991; chapter 15 of this text), Zinyama et al (1990)[2] and, indirectly, in the findings of the Poverty Assessment Survey Study undertaken in Zimbabwe in 1995 (and reported in Chapter 8 of this text).

This chapter reports on research into what urban migrants in Zimbabwe think about land reform. Everybody in Zimbabwe has a stake in the Land Reform Programme in some sense, since it is part of government expenditure, and clearly different groups vary in what they stand to gain or lose. The government and the commercial farming sector can easily make their views known, but the views of the others with less power are less easily promoted or uncovered. The evaluations reported here come from low-income African Zimbabweans who have not been resettled and whose views on land reform were deliberately sought. This latter point is important since in the national poverty survey in 1995 few respondents specified 'land' as an issue that contributed to poverty. This could be interpreted as meaning that land redistribution does not contribute to poverty alleviation. However, firstly, the questions asked in the survey were very simplistic and could have elicited responses relating to respondents' perceptions of the day-to-day causes of poverty (e.g. poor pay, no job), rather than the more indirect, contextual causes. Secondly, the vast majority of answers could be summed up in a rather unilluminating category of response: 'poverty is caused by not having enough money'. A clear indicator of this weakness is that equally few people identified lack of education as a cause of poverty. This does not mean that the alleviation of poverty could not be contributed to by improving access to education. Similarly, chosing factors other than 'the provision of land' does not mean that improving access to land cannot also help alleviate poverty in Zimbabwe.

The views of low-income African Zimbabweans on land reform are significant because ultimately the success or otherwise of a policy should be judged by the people it affects, and not only by those implementing it from above, or external 'objective' observers. Furthermore, if widespread support for a policy is evident amongst the population, this provides an excellent argument for the policy to be maintained or strengthened; and if problems are identified by the people these should be taken into account in future policy. One of the main purposes of this chapter therefore is to let the 'voices' of local people be heard.

Whose Voices?

The 'voices' reported on here belong to a large sample of adult migrants who had recently come to live in low-income areas of Harare, who were surveyed in 1994.[3] The respondents of this survey were expected to have interesting and informed views on land reform since most rural-urban migrants have an active interest in rural development given the vital nature of rural-urban linkages for many (Potts and Mutambirwa, 1990; Potts,

1996). Furthermore there are issues about the resettlement programme which very specifically affect migrants who become urban workers.

The migrants' perceptions about the resettlement programme were gathered via semi-structured interviews. The choice of topics discussed or mentioned was therefore entirely theirs. Thus it is believed the survey is a reasonably true reflection of the people's 'voices'.

People's Perceptions of the Resettlement Programme

The resettlement programme in Zimbabwe has, in many ways, had a 'bad press'. This is partly due to the effectiveness of the various lobbies which have vested interests in maintaining an unequal distribution of land and partly because these have been operating in a favourable ideological climate, for the 1980s and 1990s have seen the ascendancy of market forces as the favoured principle for allocating national and global resources. It is perhaps this ideological climate which prevented the British government, the most important donor for the Land Reform Programme, from responding positively to its own (Overseas Development Administration) evaluation of the resettlement programme in 1988. This report showed that the programme was one of the most successful planned developments in Africa, had brought considerable benefits to the majority of resettled families, and yielded an impressive economic return (estimated by the Economist at 21 per cent) (Cusworth and Walker, 1988; Palmer, 1990; Chapters 3 and 4 of this text). Yet the British government has maintained a generally negative stance towards the programme, and has done nothing to publicise this objective evaluation, whilst taking every opportunity to support the Commercial Farmers' Union line that nothing must be done to damage the (large-scale) commercial farmers of Zimbabwe. Policies predicated to any extent on principles of equity, justice and the alleviation of poverty do not fare well in such a climate.

Even those who were, or are, in favour of land redistribution for equity reasons, many of whom would also judge that national agricultural 'efficiency' would not thereby be much affected (or possibly enhanced), have also been trenchant critics (e.g. Cliffe, 1988; Palmer and Birch, 1992; Alexander, 1993). A vast range of problems has been identified, many of which are discussed in other chapters in this text. All this can make for rather depressing reading on Zimbabwean resettlement.

Given this 'bad press' there was some expectation that this survey would uncover similarly gloomy and cynical perceptions of land resettlement amongst the respondents. This did not, however, prove to be the case - as shown in Table 10.1, the people interviewed were overwhelmingly positive about the programme. This simple fact implies a major degree of success for the programme, in terms of those interviewed being positive about it, and the value of expanding it, to enhance and maintain this. Those with purely negative or neutral views were in a very small minority - only 4 per cent of the sample.

The Benefits of Resettlement: Land, Crops and Living Standards

Nearly four-fifths of the interviewees expressed a positive opinion about resettlement (see Table 10.1), saying that the programme was 'good', and most then went on to identify and describe positive aspects (Table 10.2). Broadly speaking these generally involved opinions about the quantity or quality of land made available, and the quantity and quality of crops that could be grown on the schemes.

Table 10.1 Migrants' perceptions of Zimbabwe's Land Resettlement Programme

Type of Comment	Per cent[a]	Number
Positive impact	79	147
of which caveat/suggestion included	24	45
Policy suggestion only	14	26
Has not improved situation in CAs[b]	1	2
Negative	3	6

a   186 respondents (of 269 in total) gave specific explanations about their views on resettlement; percentages are out of these 186 comments.
b   CA (Communal area).

Many people emphasised that resettlement had improved access to land for small-scale African farmers. This is, however, a very general benefit. Disaggregating what people actually said about access to land indicates different emphases about their perceptions of the programme. For many, the issue was simply that land resettlement gives people more land, but others specified that settlers benefited because they gained large lands.

Given that a family on the most common scheme of resettlement (Model A) receives 5 hectares (c. 12 acres) of arable land, plus grazing for up to 20 livestock units - which is very definitely more than the average amount available to the poorer households in most communal areas - it is not surprising that some respondents perceived that settlers' landholdings were large.

Table 10.2 Benefits of resettlement programmes identified by migrants

Type of comment	percentage*	Number
Positive impact	79	147
Of which caveat/suggestion included	24	45
Benefits specified:		
Improved access to land	41	77
Of which:		
Gives people more land	18	33
People get 'large' lands	11	20
People get enough land	8	14
Landless people get land	5	10
People get more grazing land	3	6
Black people own the land	2	3
Settlers can grow surplus for sale/money	15	27
Settlement land is fertile/productive	6	11
Settlers can grow all the crops they want	5	9
Settlers can improve standards of living	4	7
Settlers get 'means of production'	3	5
Settlers can grow lots of food	2	4
Settlers can get govt. assistance	2	3
Resettlement means less crowding	5	10

*   186 respondents gave specific explanations about their views on resettlement: percentages are out of these 186 comments.

Another frequent comment was that the settlers get enough land, which begs the question: enough land for what? There were variations in migrants' ideas about what settlers should be doing with their land, and ultimately, therefore, what the purpose of land resettlement should be. This is, of course, one of the key issues in the debate about land reform in Zimbabwe.

The ascendancy of market principles, or at least the imperative of maximising short-term production, has not passed Zimbabwe's resettlement programme by. For many commentators, and most of those involved in funding the programme, including to a large extent the government, resettlement must largely be judged on economic criteria. Settlers must be commercial in their orientation, and not just 'subsistent'. This is partly a response to the lobbying about the need to guard against damaging Zimbabwe's commercial agricultural output by redistributing land uneconomically or inefficiently, if at all. This has led to a marked shift in the government's selection criteria for settlers, from the most needy (e.g. the landless, the very poor, ex-combatants) to those who are deemed to have the necessary 'know how', and preferably, capital and equipment, to embark on commercial agricultural production (Moyo, 1994, 1995). Evidently this shift means that settlers must have enough land to grow crops and raise livestock both for their own consumption needs, and a marketed surplus. However, by no means all of our respondents were convinced about the importance or need for a commercial orientation.

Some respondents felt that the key aspect of improving access to land was that landless people got land: an issue which was also clearly important for the African political elite in the rhetoric of the liberation war and the early years of independence but which is now much less evident in those circles. It is noteworthy that this is still an issue which resonates with Zimbabwean people. Only three respondents made any reference in terms of improved land access to the racial roots of land inequality, by pointing out that resettlement meant that 'black people owned the land'.

The next most frequently identified benefit of resettlement, after access to land, was a 'commercial' factor - that settlers could grow a surplus for sale, and thus raise cash. The implication is that settlers are perceived to be more favourably placed in this respect than farmers in the communal areas (as clearly indicated by the data presented in chapter 9). This might give heart to those for whom the production of marketable surpluses should be a major determinant of the nature of the resettlement programme.

Several respondents believed resettlement land tended to be fertile even though most of the land allocated has been in the least favourable agroclimatic regions of the country (Regions IV and V). It may be that these respondents knew of particular instances where settlers had been fortunate in the farms where they had been resettled or that, as noted by Zinyama *et al* (1990), they were influenced by the fact that in the early years resettlement farmers sometimes benefit from 'stored up' fertility in formerly white-owned farms where the land had been underutilised for decades.

Other positive perceptions were that settlers could satisfy their needs in terms of the range of crops they could now grow, or in terms of the amount of food they could grow. Some felt that settlers were able to achieve better living standards than they would have had before. Other benefits specified included the fact that settlers obtain the 'means of production'; that they get government assistance; and that resettlement means less crowding in the communal areas.

Improving Zimbabwe's Resettlement Programme

Many interviewees made sugggestions about how the Land Resettlement Programme could be improved, or where it had gone wrong (Table 10.3). It is important to understand that the majority of these criticisms or suggestions were made in terms of a rider to the generally positive responses to resettlement discussed in the preceding section, although only some offered suggestions for how the programme might be improved (Table 10.1).

Three issues surfaced as being uppermost in the respondents' minds. The first was that the selection procedures were unfairly discriminatory, particularly against 'workers'. There was an early prohibition on households with a wage worker gaining access to resettlement land. Although there was a certain logic in this approach, in terms of selecting the most needy (refer to Chapter 8 for levels of poverty by land-use sector), there are clearly problems too. For example, one of the reasons why people migrate to urban areas is that they are landless, and indeed 62 per cent of those interviewed fell into this category. Some of the young migrants may have the expectation of obtaining land from their parents at some later stage - but this applies nonetheless. Some of the respondents were particularly opposed to this aspect of the schemes, and it seems that many of them had some interest in resettlement for themselves (although very few had actually ever applied).

The next two issues identified by the migrants reflect, to a large extent, two directly opposing opinions about what land redistribution in Zimbabwe should try to achieve - differences of opinion which accurately reflect certain aspects of the resettlement debate in academic and policy-making circles.

Table 10.3 Problems and policy suggestions identifed by migrants relating to the resettlement programme

Type of comment *Main problems and policies identified:*	*Percentage*	*Number*
Everyone must be eligible (e.g. workers too)	6[a] (17)[b]	12
Must allow 'communal' farming / not dictate practices	6 (15)	11
Must educate farmers and/or choose those with skills	6 (15)	11
Settlers must be given title deeds	4 (11)	8
Need more infrastructure	3 (7)	5
Need more equipment	2 (6)	4
Need more capital	2 (4)	3
Resettlement is causing environmental problems	2 (6)	4
Elites/politicians taking resettlement farms	3 (8)	6
Must settle 'good' natural farming regions	2 (4)	3

a   186 respondents gave specific explanations about their views on resettlement: unbracketed percentages are out of these 186 comments.
b   72 respondents gave explanations which identified or included a problem which they felt had either been caused by the resettlement programme, or which needed to be addressed for the programme to be successful: the bracketed percentages are out of these 72 comments.

The first point was that the settlers should be allowed to themselves choose their farming practices and lifestyles. Some respondents felt very strongly that the degree of interference in this regard by programme implementers was completely unacceptable, because redistribution of land was about redressing past inequities in access to land, and not about improving agricultural production.

The insistence on dictating how and what people should grow, setting marketing objectives, restricting livestock numbers and grazing practices - all of which are commonplace on resettlement schemes as the government attempts to assuage critics who predict that land redistribution will lead to economic and environmental disaster - was evidently perceived by some respondents as quite unacceptable. The flavour of people's feelings about this issue become clearer from their own comments, some of which are reproduced below:

- People should live near their fields, with no limit on the number of livestock;
- Government must not interfere with the life of resettled people;
- People must be allowed to grow what they want, not be dictated to;
- The government must not dictate what is grown;
- Farmers must be allowed to lead a communal way of life;
- Farmers must be given more autonomy;
- Farmers must grow what they want and there must be communal ownership of the land;
- Settlers must be allowed to live life similar to that in rural areas.

On the other hand, there were those who felt that the schemes should 'educate' farmers about how to farm, or should select settlers who already had appropriate skills. Clearly these migrants had a rather different view of the purpose of resettlement than the previous group, and were far more accepting of the argument that land in Zimbabwe must be used 'productively' and that it is therefore reasonable to discriminate in favour of those who have proven farming skills. Again their own comments are illustrative:

- Settlers must be educated on good farming methods;
- Willing and good farmers should be settled;
- Farmers must be given skills or else they will sell government inputs;
- Good farmers must be settled, lazy ones will sell fertilisers;
- People with farming knowledge must be preferred.

Although these comments indicate that this group of respondents was concerned that resettlement land should be used 'skillfully', it is notable that none specified that settlers should be made to produce for the market. For some this may have been the sub-text of their comments, but there can be no certainty that subsistence farming, as long as it was done skilfully, would have been regarded by all these people as wasting the resource of redistributed land. It is my view, however, that this is the view of most of those making relevant decisions in the current Ministry of Agriculture, as well as the main donors to the resettlement programme.

Currently settlers have a lease on their land, which, in theory, can be revoked should they not comply with the various terms and conditions of the scheme. This is the most significant 'stick' that government representatives have to make settlers practice the sort of agriculture deemed necessary for the scheme. The interviewees who mentioned this were all strongly opposed to this system: they felt that the settlers should all be given their title deeds so that their right to the land was assured. The importance of land tenure is also raised in Chapter 6 of this text.

One of the most contentious and high profile elements of the resettlement debate in Zimbabwe is that settlers are causing (or will cause) environmental problems by using inappropriate farming techniques. Yet few of our respondents felt that this was a problem and their opinions were to some extent balanced by a small group who identified another issue central to these debates about relative productivity and environmental impacts - that only a minority of resettlement land is located in the country's best agroclimatic areas (Natural Regions I and II), whilst most is in areas marginally suited, or not suited to rain-fed arable agriculture (Natural Regions III and IV). As one person stated: 'People must be settled in regions I, II and III and not in drought-stricken IV and V' (for further explanation of these natural regions, refer to Chapter 5).

A final problem identified by our sample was an issue which we had anticipated might be uppermost in their minds, given the timing of the survey: the fact that various members of the black elite had just been 'caught' leasing farms which had ostensibly been bought for resettlement purposes. Yet only six migrants mentioned this problem.

Discussion

The most important outcome of the survey reported here is the extremely high level of support and positive feelings about the programme displayed by our sample. The range of opinions expressed amongst the rural-urban migrants surveyed indicated particular emphases in their perceptions too, which are briefly reviewed below:

- Those interviewed were very concerned about the discriminatory selection procedures which many felt to be unfair and not in the spirit of what they feel resettlement should be addressing.
- Strong opposition to the ways in which the government is trying to control the nature of settlers' agricultural practices was also detected - opposition which resonates with the long history of Zimbabwean peasant struggles against attempts to control and direct their agriculture (e.g. see Drinkwater, 1991; Alexander, 1993; Chapter 11 of this text).
- The fact that settlers only hold a lease on their land was clearly deemed to be unfair.
- It is arguable that a strong element of 'moral economy' issues are influencing feelings about the resettlement programme. The nature of many of the comments fit in with the view that within Zimbabwe the right to land for Africans is (or should be) inalienable. In that context, support for land redistribution from the former white commercial farming areas means support for restoring such land rights, without conditions being attached, and thus moral or economic 'worth' should not affect one's right to land.
- In relation to the thorny issue of productivity on resettlement schemes, whilst there was opposition to government interference and support for the right to act as 'communal' farmers, this does not mean people want to be 'subsistence' farmers and only produce for themselves. First, many communal farmers produce for the market and second, the benefit most frequently specified (besides the general one that more land was being made available), was that resettlement schemes enable beneficiaries to produce surpluses.
- There was a minority perceptions that settlers should be educated farmers, indicating that the same tensions which exist in the land reform debate in scholarly and political circles can be found at the 'grass roots'. However, whilst most of those surveyed had come from the rural areas, over half of those making this point (55 per cent) were urban-urban migrants, so their direct experience of farming was more distant or non-existent. Thus their ability to balance negative media reports about 'uneducated' resettlement farmers against knowledge of

the real constraints faced by peasant farmers in Zimbabwe would be, on average, less than for the sample as a whole.

- Much research in the 1980s and 1990s has indicated that the causes and rate of environmental changes in the communal areas have often been exaggerated or misunderstood (e.g. Biot *et al*, 1992; Wilson, 1990; Elliott, 1989; Macgregor, 1991; Scoones, 1990; Cousins, 1989; Chapter 12 of this text), and even those who maintain that Zimbabwean peasants cause environmental degradation often simultaneously admit that the major reason is that the communal areas suffer unduly high population pressure due to the history of unequal land distribution (e.g. Whitlow, 1988). The responses in this survey may suggest that rural-urban migrants are generally not very convinced that environmental degradation is the problem it is sometimes made out to be, perhaps because they have more direct experience of rural environmental conditions. Alternatively, or in addition, resettlement may be seen as a way of escaping the environmental problems of the communal areas. It is arguable that the frequent reference to settlers obtaining 'enough' land or large lands, that it means less crowding, and that settlement land is fertile, all point in that direction - and furthermore indicate that the respondents do not assume (as do some of the opponents of the programme) that settlers, farming at much lower densities, will face the same sort of environmental dilemmas that many communal area farmers must deal with.

- Another issue raised by the survey was how resettlement affects the communal areas themselves. Views on this varied quite widely (see Potts and Mutambirwa, 1997 for more details) but given the limitations of the programme for improving the situation in the communal areas, the fact that nearly a third of those interviewed were generally positive about the impact of resettlement in their communal area, could be viewed as a reasonable success.

Conclusions

This chapter shows how a group of low-income urban Zimbabweans, mostly having strong rural connections, perceive the Land Reform Programme. Their views were variable, showing an awareness of most of the issues central to debates about land redistribution in Zimbabwe in official and academic circles. However, the significance attached to these issues by urban migrants varies significantly from the weight given to them in these other circles. In broad terms, matters of justice and equity were deemed far more important than economic criteria, although the opportunities afforded by resettlement for commercial production were praised by many respondents. Amongst the respondents there remains a strong affection for, and attachment to, the concept that land rights should come without conditions.

In general the voices of the people from this survey were loud in their praise for the programme, which was perceived to deliver many benefits. This support is reflected in the opinions of those respondents (see Table 10.3) who demanded that resettlement be speeded up, because, as one of them put it, 'people scratching on barren land are still waiting for resettlement'. For the people of Zimbabwe then, land reform, despite its many problems, has been a success and it is to be hoped that donors and policy makers will listen to their 'voices'.

Notes

1   Coverage of settlers' and peasants' views in local magazines such as *Moto* and *Parade* has been much more frequent.
2   See Potts and Mutambirwa (1997) for more details on these studies.
3   This research was part of a long-term collaborative project on migrants and migration to Harare begun in 1985. The details of the survey methodology are contained in Potts and Mutambirwa (1997).

References

Alexander, J. (1993), 'State, peasantry and resettlement in Zimbabwe', *Review of African Political Economy*, 61, pp. 325-45.

Berry, B.B. (1991), *The impact of agricultural resettlement in Zimbabwe: the Soti Source Model A intensive resettlement scheme*, MA thesis, University of Witwatersrand.

Biot, Y, Lambert, R. and Perkins, S. (1992), *What's the problem? An essay on land degradation, science and development in sub-Saharan Africa*, (Discussion paper no. 222), School of Development Studies, University of East Anglia, Norwich.

Cliffe, L. (1988), 'Zimbabwe's agricultural 'success' and food security in Southern Africa', *Review of African Political Economy*, 43, pp. 4-25.

Cousins, B. (ed.) (1989), *People, land and livestock: proceedings of a workshop on the socio-economic dimensions of livestock production in the communal lands of Zimbabwe*, Centre of Applied Social Science, University of Zimbabwe, Harare.

Cusworth, J. and Walker, J. (1988*)*, *Land Resettlement in Zimbabwe: a preliminary evaluation*, ODA (Evaluation Report EV 434), London.

Drinkwater, M. (1991), *The state and agrarian change in Zimbabwe's communal areas*, London, Macmillan.

Elliott, J. (1989), *Soil erosion and conservation in Zimbabawe: political economy and environment*, Ph.D. thesis, University of Loughborough.

Jacobs, S. (1991), 'Changing gender relations in Zimbabwe: the case of individual family resettlement areas', in D. Elson (ed), *Male bias in the development process*, Manchester University Press, Manchester.

Kinsey, B.H. (1984), 'Resettlement: the settlers' view', *Social Change and Development*, 7.

Kinsey, B.H. (1999), 'Land Reform, Growth and Equity: Emerging Evidence from Zimbabwe's Resettlement Programme', *Journal of Southern African Studies*, 25, 2, pp.169-192.

Macgregor, J. (1991), *Woodland resources, ecology, policy and ideology: an historical case study of woodland use in Shurugwi communal area, Zimbabwe*, unpublished Ph.D. thesis, University of Loughborough.

Moyo, S. (1995), *The land question in Zimbabwe*, SAPES, Harare.

Moyo, S. (1994), *Economic nationalism and land reform in Zimbabwe*, SAPES, Harare.

Palmer, R. and Birch, I. (1992), *Zimbabwe: a land divided*, Oxfam, Oxford.

Potts, D. and Mutambirwa, C.C. (1990), 'Rural-Urban Linkages in Contemporary Harare: Why Migrants Need Their Land', *Journal of Southern African Studies*, 16, 4, pp. 676-98

Potts, D. (1996), 'Zimbabwean migrants must keep their land: support for the Land Tenure Commissioner's Report', *Zimbabwe Review*, September.

Potts, D. and Mutambirwa, C. (1997), '"The government must not dictate...": Rural-urban migrants' perceptions of Zimbabwe's land resettlement programme'. *Review of African Political Economy*, 24, pp. 549-66.

Scoones, I. (1990), *Livestock populations and household economics: a case study from southern Zimbabwe*, Ph.D. thesis, Imperial College, University of London.

Whitlow, R. (1988), 'Soil erosion and conservation policy in Zimbabwe: past, present and future', *Land Use Policy*, pp. 419-33.

Wilson, K. (1990), *Ecological dynamics and human welfare in Southern Zimbabwe*, Ph.D. thesis, University College London.

Zinyama, L., Campbell, D. J. and Matiza, T. (1990), 'Land policy and access to land in Zimbabwe: the Dewure resettlement scheme', *Geoforum*, 31, 3, pp. 359-70.

11 The Enduring Appeal of 'Technical Development' in Zimbabwe's Agrarian Reform

DR J. ALEXANDER

Introduction

Highly intrusive state intervention into the ways in which Africans live and farm has long been a defining feature of agrarian reform in colonial and post-colonial Zimbabwe. The 'technical development' policies of the 1950s marked the consolidation and extension of earlier reforms and cast them into a much broader programme of transformation. These interventions are remembered perhaps above all for the violent resistance they sparked, and the boost they gave to nationalist mobilisation. Less well-remembered is the concerted critique they provoked from within the Rhodesian state's own technical departments. After independence, continuities with the technical development policies of the 1950s were pronounced, both within the communal areas, and on the new resettlement schemes. As in earlier periods, the interventions of the 1980s and 1990s provoked popular resistance as well as wide-ranging criticism from the state's experts and academics.

This chapter seeks to ask why technical development has had such an enduring appeal. We must consider not only how economic and political forces have influenced agrarian reform, but also how institutional and ideological factors have come into play.

Colonial Intervention: The Rise and Fall of Technical Development

The Rhodesian state's regulation of African lives was far from consistent in theory or practice. At first, 'native policy' envisaged a move towards 'detribalisation' through proletarianisation (Steele, 1972). Later, Africans were viewed as living in a society apart, bound by customs and traditions opaque to Europeans, save for the experts of the Native Affairs Department (NAD). Alongside its traditionalist views, the NAD put forward a set of policies that would later be termed 'technical development', encompassing demonstration, centralisation, and destocking, and culminating in the Native Land Husbandry Act of 1951. They were justified in very different ways, but all were premised on beliefs that assumed the superiority of western culture and science. Africans, of course, were judged not to have science (see Beinart, 1984).

Early interventions combined a concern for 'civilising' and evangelising alongside the promotion of 'modern scientific agriculture', meaning intensive, permanent cultivation, crop rotation and the use of fertiliser and improved seeds (Drinkwater, 1989; McGregor, 1991). These demonstration policies were followed by the more intrusive policy of centralisation, promoted for its aesthetic and administrative benefits as well as promises of increased productivity. The reserves were to be 'rationalised' and 'enframed' into discrete, standardised compartments so as to produce ordered, neat colonial subjects (Werbner, 1992). During the 1930s depression, emphasis was laid on centralisation's capacity to fit more Africans into reserves and promote conservation (Palmer, 1977; Steele, 1972).

In the 1940s, these interventions were superseded by a more coercive and interventionist ethic, promoted by a rapidly growing corps of technical officials within the NAD, and justified by a region-wide concern for conservation (Beinart, 1984). Alarm over 'overstocking' led to the passage of compulsory destocking regulations in 1943. More and more officials entered the reserves, armed with an ever-growing array of coercive powers, and intent on regulating all aspects of African agriculture. They became, as one Native Commissioner put it, 'virtual policemen', bent on prosecuting 'agricultural crimes' (National Archives of Zimbabwe, 1947). African farming methods were, in effect, criminalised (Wilson, 1986).

But this was merely the presage to an even more ambitious state-directed plan for intervention. In 1951, the Native Land Husbandry Act (NLHA) was promulgated amidst an economic boom and great confidence in the prospects for state-planned modernisation. Labour shortages and pressure on the reserves due to massive evictions from 'white' land confirmed the need in the official mind for a radical restructuring of African participation in the colonial economy. The passage of the NLHA was preceded by a reorganisation of the NAD that gave the initiative to the technical branches, as opposed to the 'conservative' administrators with their traditionalist views (Holleman, 1969).

The NLHA reproduced the earlier policies of centralisation, destocking, conservation and agricultural intensification, but added to these the radical goal of ending labour migration between reserves and work places, and halting further settlement in the reserves by issuing saleable land and stock rights to a permanently limited number of African farmers. Africans were to become full time workers or farmers, and tenure was to be individual. Officials expected an 'almost immediate beneficial result' in productivity and conservation to result from the NLHA's secure tenure (National Archives of Zimbabwe, 1958; Phimister, 1993).

The Act altered the nature of colonial subjectivity: it directly repudiated 'customary' and 'communal' rights to land in favour of individual right holders and 'secular state power', i.e. the government officials charged with monitoring land use and transfers (Wilson, 1987). Official rhetoric emphasised the new role of the marketplace in regulating access to land, but in fact a host of restrictions governed eligibility for land and stock rights, the amount of land and stock owned, and the accumulation of stock and land. Land could not be used as collateral against a loan. African 'rights' to land were subordinated to the imperatives of conservationist concerns and the limits of territorial segregation (see Chanock, 1991).

These regulations reflected officials' belief in the need for state control over the process through which Africans' 'backward and inefficient' farming methods were to be modernised. If officials were willing to concede that Africans could be modernised, they were not willing to let them do it themselves. African farmers were excluded from the planning process by supposedly apolitical technocrats, acting in their best interests. There followed an immensely expensive programme of collection of technical data, censuses, assessments of carrying capacities and soil types, so as to allow the state to intervene, and to justify and enforce its plans. Technical officials exercised unprecedented control, driving on the NLHA's implementation with alarmist assessments of conservation in the reserves.

But colonial science proved far from reliable, as contemporary officials and academics have noted at length (Alexander, forthcoming). Projections of productivity increases proved ill-founded; the 'economic units' proved unachievable and there were serious errors in the implementation process. There simply was not enough land for all eligible farmers, nor enough stock for them to maintain fertility in their fields. Security of tenure was greatly undermined, not strengthened. There were too few jobs for those excluded from land, and inadequate provision for the settlement of a permanent working class in towns. Moreover, the extent to which agricultural productivity relied on migrant earnings had been sorely underestimated.

These problems were heatedly debated by officials. Native Commissioners focused on the political discord the Act left in its wake, but there were also criticisms from technical officials, especially economists. NLHA implementation was suspended on irrigation schemes and high rainfall areas due to new economic evaluations, and wider doubts about the NLHA model as a whole were underlined by the findings of a survey undertaken in 1959/60 which showed that the NLHA had had negative effects on productivity in most regions (Holleman, 1969).

The cutting criticisms of the NLHA's most basic assumptions from within the technical cadres which had so recently championed the goals of 'modernisation' did not, however, stand as the lasting explanation for the demise of the NLHA. Blame for the NLHA's failures was rapidly shifted from its own flaws to the nature of African society (Alexander, forthcoming). Africans simply were not ready for modernisation: they were unable to grasp the rationality of science and the market. The Rhodesian state retreated into its management of Africans as 'tribal communities', subject to the rule of chiefs and headmen, maintaining only its commitment to conservationist measures in the turbulent years of the 1960s and 1970s. But the sciences of technical development would survive to see another day.

Post-Independence Technical Development

Technical development is a less obvious legacy of the Rhodesian era than the unequal division of land between black and white. It was profound in its effects nonetheless (Drinkwater, 1988, 1989). In the early 1980s, the new ZANU (PF) government introduced significant change: it expanded services to black farmers in the former reserves, discriminatory legislation was repealed and an ambitious programme of land redistribution through resettlement was tabled. The new government did not, however, challenge the beliefs and practices that had informed technical development – what Drinkwater calls the state's 'purposive rationality'. The powerful, inherited technical ministries remained beyond the authority of the newly established hierarchy of rural democratic institutions, the district councils and village development committees, and the remaining (now marginalised) 'traditional' hierarchy of chiefs, headmen and village heads.

In the early 1980s, land redistribution was cast as a political imperative above all else, necessary to control popular expectations, and to create stability (MEPD, 1981). Few questioned the political importance of redistribution, but it was near impossible to find voices that questioned the old, historically unfounded assumptions about the inherent environmental destructiveness and lack of productivity of African farmers, as opposed to their white counterparts (see Palmer, 1977). Technical development assumptions, along with their authoritarian baggage, ran through many early documents (Ranger, 1988; Drinkwater, 1988; Alexander, 1994).

Policy-makers and the many official commissions relied on the same bodies of research and ideas about the tenure and productivity of the former reserves, now communal areas, as those that had informed Rhodesian policy. Early government documents held that the answer to 'bad land distribution, poor soils, poor farming methods and over-population' was to be found in resettlement and the inculcation of 'good farming methods', identified through 'considerable investigation, research and planning, through the employment of special expertise and skills'. These plans would need to be 'adequately sold to the people' such that they were convinced these were 'in their best interests' (MEPD, 1981). Repeating a familiar formula, the 1981 Riddell Commission (1981) recommended consolidating arable land into blocks, fencing grazing areas, registering land with a title, and abolishing labour migration, thus creating permanent farmer and worker populations. The 1982 Chavanduka Commission blamed low communal area productivity on 'traditional tenure, poor farming practices and labour migration'. It recommended greater security of tenure 'recognisable in practice and law', i.e. dependent on state intervention and regulation (1982).

The tendency to draw on colonial ideas and practices did not diminish with time. The Ministry of Lands' *Communal Land Development Plan* (1985) relied on 1970s research and reports, ignoring the increased communal area contribution to marketed crops of the post-independence years. It cast communal area tenure as 'traditional', and hence static, conservative and opposed to accumulation and production for the market. In fact communal tenure had been anything but static and conservative, and it had proved far more secure than tenure dependent on the state. Assumptions as unreasonably optimistic as those of the 1950s were made about the possibility of ending labour migration, and little recognition was given to the interdependence of rural production and urban earnings (see, *inter alia*, Bush and Cliffe, 1984; Drinkwater, 1988; Ranger, 1988; Scoones and Wilson, 1988; Cousins, 1990). The plans and analyses presumed a particular relationship between rural people and the state: it was authoritarian, top-down, based on the state's superior mastery of the practices, ideas and information that informed technical development. Despite the many dramatic changes of the 1980s, continuities with Rhodesian planning remained rooted in the reproduction of the same ideological assumptions and bodies of expertise within bureaucracies that remained unaccountable to representative institutions.

The resettlement programme provides a stark illustration. The ministries concerned with resettlement stressed that the measure of the programme's success lay in its ability to produce marketed surpluses, an ability that they firmly believed did not exist within the 'subsistence-oriented' communal areas. Emphasis was laid on the need to redistribute land on a 'planned and organised basis', so as to 'create an agricultural community on land which will no longer be just subsistence but commercial in orientation'. Resettlement could not become 'just an extension of a peasant sector or subsistence farming sector' (MEPD, 1981). Resettlement schemes were envisaged as self-contained islands of modernisation. No 'reversion back to traditional methods of agriculture' could be allowed (see Drinkwater, 1988). Embedded in this formulation was a rejection of historically based claims to land, i.e., claims that did not rely on the state's technical assessments, plans and allocations.

Unsurprisingly, the dominant resettlement model, Model A, closely followed the pattern of the NLHA. Settlers were allocated individual, demarcated plots of arable land with provision for communal grazing dependent on the agro-ecological region. Settlers' access to homes, land and grazing was dependent on a series of permits, issued by the Department of Rural Development (Kinsey, 1983).

From the mid-1980s, the emphasis of agrarian policy shifted to improvements within the communal areas, not land distribution. The political and economic reasons for this shift, though still much debated, need not be rehearsed here (see, *inter alia*, Cliffe, 1986; Moyo, 1986; Bratton, 1987; Weiner, 1989; Palmer, 1990; Herbst, 1990; Dashwood, 1996; Moyo, 2000). Central to justifying this shift in policy were the following assumptions. Firstly, it was assumed that communal area problems could not be solved by continued land redistribution - to do so would mean threatening the productive sector of the commercial (formerly white) farms. Secondly, it was assumed that communal areas could be made more productive through the improvement of land use therein.

In the latter half of the 1980s, government ministries managed agrarian reform with little reference to the people for whom plans were developed.

Debates focused on methods and obstacles to implementation, not on the merits of state-regulated modernisation. As with the NLHA, however, these bold plans for change often ran aground on bureaucratic conflict and local resistance.

In the government's *First Five-Year National Development Plan*, published in 1986, resettlement came to mean two things: 'translocation resettlement', meaning what had simply been called resettlement previously, and 'internal resettlement', meaning reorganisation within the communal areas. A full 20,000 families were to be 'internally resettled' annually. They were to be moved into 'consolidated villages', arable land was to be planned in blocks, and rotational grazing schemes and irrigation were to be introduced (Government of Zimbabwe, 1986). The link between internal and translocation resettlement was weak: there was no coordination between the implementing agencies for land acquisition and communal area reorganisation. Ministerial reshuffles in 1985 deepened the difficulty by dividing responsibility for resettlement between two powerful ministries - Local Government and Lands (Alexander, 1994).

Oddly, it was a third Ministry - Public Construction and National Housing - which took the initiative, making use of a budgetary allocation for a rural housing programme. The 'consolidated village', in which the improved houses were to be built, was the lowest rung of a seven-tier settlement hierarchy which drew on earlier traditions of physical planning. The hierarchy appealed as a piece of social engineering: it offered the promise of rapid and 'visible results of "development" to rural areas', and 'a single and well-defined social tree - orderly, controlled, simple' (Gasper, 1988). A 1986 directive required a 'pilot village' in each district. Little consultation took place; people were simply expected to move to the new sites without compensation and to take out loans to pay for the new houses. The unsurprising lack of response was met with criticism of district councils for their failure to 'make people understand and accept the planned villages' (Association of District Councils, 1986).

Despite being a vivid example of the problems of top-down planning, villagisation remained firmly on the agenda, now as a component of the broader agrarian reform policies of the Ministries of Local Government and Lands. These policies paralleled with remarkable closeness the thinking that had informed the NLHA, as well as the early 1980s commissions on land reform. Once again, emphasis was placed on state regulation of communal area farming - the extension of resettlement area-style permits into the communal areas was even mooted, so as to 'guarantee succession, prevention of sub-division, abidance by environmental regulations and acceptable land husbandry practices'. Those who failed to comply would have their rights removed (Nyanga Symposium, 1987). The technical assessments that would provide the basis for intervention were to be provided by the Ministry of Lands' Agritex, in the form of land-use plans (Gonese, 1988).

These new initiatives emphasised the need to grant further powers to the elected village development committees to allow them to enforce penalties for failures to farm productively, and to co-opt traditional authorities. These changes were explicitly designed to allow more effective state intervention – not more effective representation. The Director of the Department of Rural Development described the problem as one in which the 'communal lands people have got rights which don't derive from government legislation. They have always occupied that land, it is theirs, it is not state land'.

Despite the confidence with which some officials promoted communal area reorganisation, there were critical views. Administrators were often reluctant to back such unpopular and disruptive plans. Agritex, though confident in the technical merit of land-use planning, did not think that the communal areas could be depopulated and destocked to the extent that their calculations required. Agritex resented the domination of their time by land-use planning, rather than extension duties, and questioned the ability of local institutions to enforce plans (Agritex, 1988). In 1994, the official Land Tenure Commission identified a range of familiar problems with land-use planning: the 'over-centralisation of government with the relevant technical ministries using top-down methods of planning and implementation', high short-term costs and risks, and poor coordination among ministries. There were few incentives for communal area farmers to participate, and sound reasons for them to resist. The Commission recommended that the link between reorganisation within the communal areas and resettlement outside them be strengthened, and that planning be decentralised (LTC, 1994).

These recommendations did not, however, filter down to district levels where a combination of donor interest in land-use planning, and the longstanding commitment to technical development among government officials remained. Councils sometimes backed the plans in order to gain access to donor funds, or because they saw the land-use plans as a means of delivering development, modernity and order, or performing other policing functions like the control of new settlement (Alexander *et al.*, 2000). These policy interventions, like those of the Rhodesian state, continued to bear little relation to the actual control of land use and tenure, though they were often invoked as a basis for disputing rights to land (Andersson, 1999). Instead, visions of agricultural production with little bearing on reality continued to inform government policy, and to justify intervention.

Conclusion

The continued commitment to technical development requires explanation. Why did such a clearly unsuccessful and unpopular policy continue to appeal? The answer is in part the result of bureaucratic continuities and the transmission of plans, information, and methods of analysis from one generation of officials to the next. The aerial photographs from the era of land husbandry, for example, were dusted off in the service of land-use planning in the 1980s. It is in part to do with the appeal of technical solutions to planners. Technical development policies offer an air of scientific detachment and professionalism - a cool, considered, apolitical analysis of a problem whose main outlines are seemingly well known. They offer order and modernity; they suit the 'purposive rationality' of bureaucracy.

We must also consider the way in which technical development assumes and seeks to create a specific kind of disciplined, dependent and ahistorical subject. Technical development assumes that communal area residents lack the necessary expertise to live and farm in a productive, conservation-conscious, modern way without state regulation. The 'petty tyranny' of these regulatory relationships is everywhere contested (Worby, 1998), but they continue to offer a means to extensive state intervention, and justification for running rough shod over historical attachments to the land, over the experience of decades of farming, and memory of resistance to previous state intervention. Like its Rhodesian predecessor, the Zimbabwean state has been divided by debate over technical development's form and implementation, but not by its value, and that is a mark not only of the enduring appeal of technical development itself, but also of a failure to construct new political relationships.

References

Agritex (1988), *The Communal Area Reorganisation Programme: Agritex Approach*, discussion paper, Agritex, Harare.
Alexander, J (1994), State, peasantry and resettlement in Zimbabwe, *Review of African Political Economy*, 61.

Alexander, J (forthcoming), 'Technical Development and the Human Factor: Sciences of Development in Rhodesia's Native Affairs Department', in S. Dubow (ed.), *Science and Society*, Manchester University Press, Manchester.

Alexander, J., McGregor, J. and Ranger, T. (2000), *Violence and Memory: One Hundred Years in the 'Dark' Forests of Matabeleland*, James Currey Pulishers, Oxford.

Andersson, J. (1999), 'The Politics of Land Scarcity: Land Disputes in Save Communal Area, Zimbabwe', *Journal of Southern African Studies*, 25, 4.

Association of District Councils (1986), Minutes of the Third Annual Congress, mimeo.

Beinart, W. (1984), 'Soil Erosion, Conservationism and Ideas about Development: A Southern African exploration, 1900-60', *Journal of Southern African Studies*, 11, 1.

Bratton, M. (1987), 'The Comrades and the Countryside: The Politics of Agrarian Reform in Zimbabwe', *World Politics*, 39, 2.

Bush, R. and Cliffe, L. (1984), 'Agrarian Policy in Migrant Labour Societies: Reform or Transformation in Zimbabwe', *Review of African Political Economy*, 29.

Chanock, M. (1991), 'Paradigms, property: A review of the customary law policies, and of land tenure', in K. Mann and R. Roberts (eds), *Law in Colonial Africa*, James Currey Publishers, London.

Chavanduka Commission, Government of Zimbabwe (1982), *Report of the Commission of Inquiry into the Agricultural Industry*, Government Printers, Harare.

Cliffe, L. (1986), *Policy Options for Agrarian Reform: A Technical Appraisal*, report submitted by the FAO for the consideration of the Government of Zimbabwe, FAO, Rome.

Cousins, B. (1990), *Property and Power in Zimbabwe's Communal Lands: Implications for Agrarian Reform in the 1990s*, Land Policy in Zimbabwe after 'Lancaster' Conference, University of Zimbabwe, Harare.

Dashwood, H. (1996), 'The Relevance of Class to the Evolution of Zimbabwe's Development Strategy, 1980-1991', *Journal of Southern African Studies*, 22, 1.

Drinkwater, M. (1988), *The State and Agrarian Change in Zimbabwe's Communal Areas: An application of critical theory*, Ph.D., University of East Anglia, Norwich.

Drinkwater, M. (1989), 'Technical development and peasant impoverishment: land-use policy in Zimbabwe's Midlands province', *Journal of Southern African Studies*, 15, 2.

Gasper, D. (1988), 'Rural Growth Points and Rural Industries in Zimbabwe: Ideologies and Policies', *Development and Change*, 19.

Gonese, F., Assistant Director, Department of Rural Development (1988), *A Framework for Communal Land Reorganisation in Zimbabwe*, Seminar on Communal Lands Reorganisation, Harare.

Government of Zimbabwe (1986), *First Five-Year National Development Plan, 1986-1990*, Vol. 1, Government Printers, Harare.

Herbst, J. (1990), *State Politics in Zimbabwe*, University of Zimbabwe Press, Harare.

Holleman, J.F. (1969), *Chief, Council and Commissioner*, Oxford University Press, London.

Kinsey, B. (1983), 'Emerging Policy Issues in Zimbabwe's Land Resettlement Programmes', *Development Policy Review*, 1.

Land Tenure Commission (1994), *Report of the Commission of Inquiry into Appropriate Agricultural Land Tenure Systems. Volume One: Main Report*, Government Printers, Harare.

McGregor, J. (1991), *Woodland Resources: Ecology, policy and ideology. An historical case study of woodland use in Shurugwi communal area*, Zimbabwe, Ph.D., Loughborough University.

Ministry of Economic Planning and Development (1981), *Zimbabwe Conference on Reconstruction and Development: Report on Conference Proceedings* (Salisbury).

Ministry of Lands, Resettlement and Rural Development (1985), *Communal Lands Development Plan: A 15 Year Development Strategy*, Government Printers, Harare.

Moyo, S. (1986), 'The Land Question', in I. Mandaza (ed.), *Zimbabwe: The Political Economy of Transition 1980-1986*, Codesria, Dakar.

Moyo, S. (1995), *The Land Question in Zimbabwe*, SAPES, Harare.

Moyo, S. (2000), 'The Political Economy of Land Acquisition and Redistribution in Zimbabwe, 1990-1999', *Journal of Southern African Studies*, 26, 1.

Nyanga Symposium (1987), *Report on the National Symposium on Agrarian Reform in Zimbabwe*, Department of Rural Development in collaboration with the FAO, Rome.

Palmer, R. (1977), *Land and Racial Domination in Rhodesia*, Heinemann, London.

Palmer, R. (1990), 'Land Reform in Zimbabwe, 1980-1990', *African Affairs*, 98.

Phimister, I. (1993), 'Rethinking the reserves: Southern Rhodesia's Land Husbandry Act reviewed', *Journal of Southern African Studies*, 19, 2.

Ranger, T. (1988), 'The Communal Areas of Zimbabwe', *Symposium on Land Reform in African Agrarian Systems*, University of Illinois, Urbana Champaign.

Riddell Commission, Government of Zimbabwe (1981), *Report of the Commission of Inquiry into Incomes, Prices and Conditions of Service*, Government Printers, Harare.

Scoones, I. and Wilson, K. (1988), 'Households, Lineage Groups and Ecological Dynamics: Issues for Livestock Research and Development in Zimbabwe's Communal Areas', in B. Cousins *et al.* (eds), *Socio-Economic Dimensions of Livestock Production in the Communal Lands of Zimbabwe*, Centre for Applied Social Science, Harare.

Steele, M.C. (1972), *The Foundations of a 'Native' Policy in Southern Rhodesia, 1923-33*, Ph.D., Simon Fraser University.

Weiner, D. (1989), 'Agricultural Restructuring in Zimbabwe and South Africa', *Development and Change*, 20, 3.

Werbner, R. (1992), *On the rationalisation of space and the critique of development*, International Conference of the Delphi Forum, Delphi, October 29-31.

Wilson, K. (1986), *History, Ecology and Conservation in Southern Zimbabwe*, Oxford.

Wilson, K. (1987), *Research on trees in the Mazhiwa and surrounding areas*, ENDA-Zimbabwe, Harare.

Worby, E. (1998), *Tyranny, Parody and Ethnic Polarity: Ritual Engagements with the State in North-western Zimbabwe*, Journal of Southern African Studies, 24, 3.

National Archives of Zimbabwe

S1563, NC's *Annual Report for 1947*, Wedza.

6.5.9R/84273, F. H. Dodd, Administrative Officer, NLHA, *The Native Land Husbandry Act, October 1958.*

12 Resource Implications of Land Resettlement in Zimbabwe: Insights from Woodland Changes

DR JENNIFER A. ELLIOTT

Introduction

There is much that is not known concerning the environmental implications of land resettlement in Zimbabwe. However, fears that land resettlement would have detrimental environmental impacts and lead to the replication of the 'communal area conditions', have been prominent since the outset of the land reform programme (Cliffe, 1988). Ten years on, the Council of Commercial Farmer's Union reported that,

> with few notable exceptions, the majority of resettlement schemes to date *have led to a serious* loss of productivity, *denudation of resources*, insufficient income and even food aid being required by settlers (1991, emphases added).

Subsequently, many Zimbabwean writers have presented resource conservation and poor peasant land husbandry as argument for the restricted pace of land redistribution and for agrarian reform in the communal areas (CAs) themselves (Moyo, 1995). In the main, however, the validity of this environmental debate to date has been limited by the general lack of objective data or systematic monitoring concerning the environmental impacts of land resettlement.

Woodland resources are known to be a central component of rural livelihoods in the African farming sector of Zimbabwe (Campbell, 1996). This chapter presents insights to aspects of the research regarding woodland changes in resettlement areas (RAs). It offers a contribution to the broader debate concerning prospective environmental changes with land resettlement. More importantly, it highlights a number of challenges for policy and the future management of woodland resources in these areas.

The 'Resource Plenty' Conditions of the Resettlement Areas under Debate

National studies of deforestation and erosion (Whitlow, 1979, 1988) found a close correlation between resource degradation and land tenure, most degradation being in the densely populated communal areas. It could therefore be expected that resettled households, on moving from the communal areas, would experience improved access to resources, including woodlands.

In 1992, research was carried out in two Model A resettlement scheme areas, namely in Wenimbi-Macheke (Natural region IIb) and Tokwe I (Natural Region III) (Elliott 1994, 1995). An explicit focus of this work was the prospective changes in the use and management of woodland resources, between the contemporary situation and household experience when resident in the communal areas. The broad patterns found (Elliott, 1994) and reported below, would suggest a general experience of improved woodland supply conditions on movement to the resettlement areas.

- Continued dependence on woodfuel sources for cooking, heating, brick- making and beer-brewing;
- An increase in total wood used;
- Increased reliance on individual rather than communal sources of wood;
- Greater active selection of species for particular activities.

In excess of 400 households were interviewed within the research. 98 per cent reported that they used solely woodfuel sources of energy for the activities included in the study (as they had done prior to movement). Generally, people assessed that they were using more wood currently and their principal supply was through the clearance of plots for cultivation. Interviewees reported that when resident in the communal areas, they had been more dependent on woodfuel collected in the common resources of the grazing areas. However, not all households were using more wood currently. In discussions concerning the species used in the varied activities, for example, it was evident that on movement to the resettlement areas, many households were able to collect particular species as desired. For 23 per cent of the sample, the availability of preferred species and the improved form of logs, were given as explanation for a reduction in total wood used over their experience in the communal areas.

A further aspect of this research involved sequential aerial photograph interpretation. Two sets of 1:25,000 monochrome contacts for eight villages within each study site, were analysed for a period spanning the designation of the area within the resettlement programme. This work revealed substantial variation in terms of the amount of woodland present at the time of designation, both between and within the schemes. For example, Tokwe I villages, on average, had over twice the proportion of land under woodland than Wenimbi-Macheke villages, as shown in Table 12.1. Particular villages were extremely resource-poor at the time at which people were resettled to the schemes. One village in Wenimbi-Macheke, for example, had only 1 per cent of the total area under woody cover prior to designation, in contrast to several villages in the Tokwe scheme, which had over 60 per cent woody cover. Furthermore, it is not possible to differentiate between particular woodland species on aerial photographs, which is known to be an important factor in the consideration of resource scarcity.

Table 12.1 Average land use and woody cover change with resettlement

	Average % area before resettlement		*Average % change*	
	Wenimbi (1981)	Tokwe (1972)	Wenimbi (to 1987)	Tokwe (to 1985)
Settlement	0.2	0.1	1.5	0.5
Cultivation	12	4.1	8	9.6
Grassland	55.5	41.8	-12	-9.5
Woodland	26.8	53.5	-0.7	-1.6

Source: Elliott, 1995

Such variation reflects not only the ecological conditions of the different regions, but also the (related) agricultural histories of the areas. For example, the relative lack of woodland resources in Wenimbi-Macheke, is in part the legacy of decision-making by former commercial tobacco growers in the area. Critically, however, there is no prior planning of woodland resources within the resettlement policy, and 'woodland management or use is not a component of the planning process' (Scoones and Matose, 1993). The planning conventions used for the resettlement policy, to date, have been heavily influenced by the practices of the commercial farming sector. In consequence, soil, water, and pasture conservation are emphasised, but there is no account given to the role of the woody biomass element, which is known to be central within small-scale African farming systems.

In summary, it cannot be assumed uncritically that all households, on movement within the resettlement programme, experience necessarily an improvement in access to woodland resources.

Patterns of Woodland Change

Sequential aerial photograph analysis has also provided the principal insights, to date, concerning the change in the overall extent of woodland cover during the first years of the resettlement programme. Grundy *et al* (1993), for example, found a 25 per cent decline in 'dense woodland' cover in Mutanda Resettlement Area between 1980 and 1986, as assessed from aerial photography. Cultivation, which had increased from less than 5 per cent of the total area prior to implementation of the programme to 35 per cent in 1986, was considered the chief source of the deforestation. Questionnaire surveys and field observations were further elements of the research which aimed to quantify the ongoing woodland resource needs in the scheme. It was concluded that contemporary levels of use could be sustained if no further extension of arable lands occurred.

This level of change was greatly in excess of the findings in Wenimbi-Macheke and Tokwe I scheme areas, as shown in Table 12.1. The major land use change identified in these cases was the decline of the grassland area (with significant differences between the two study sites). The contrast with the Mutanda scheme may be explained by the greater levels of cultivation evident at the time of designation, particularly in Wenimbi-Macheke. It is likely that such crude categorisations of land use may mask important insights to the patterns and directions of woodland change. Woodlands are a renewable resource, such that as well as woodland losses, there may be significant gains occurring at specific places and points in time. Figure 12.3 describes the land use changes in one village in the Tokwe I scheme, Devon Ranch, between 1972 and 1985. The data has been obtained through aerial photograph interpretation and digitised within a Geographic Information Systems (GIS) package (ArcView).

1972

1985

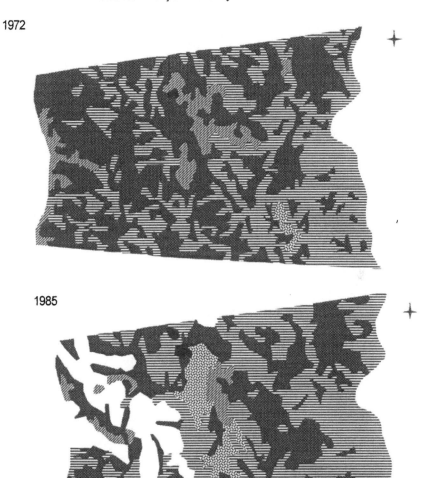

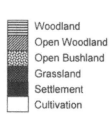

Woodland
Open Woodland
Open Bushland
Grassland
Settlement
Cultivation

Figure 12.1 Land use change in Devon Ranch, Tokwe I, 1972-85, mapped through Arc View

Source: Elliot, 1995

Woodland resources have been differentiated by height of trees and density of cover into four distinct categories. 'Woodland' was defined as those areas which had in excess of 50 per cent canopy; 'open woodland' as 10 to 50 per cent canopy; 'bushland' as more than 50 per cent canopy but distinguished from woodland by the height of trees (smaller); and 'open bushland' was identified as similar areas of bush and scrubland but with between 10 and 50 per cent canopy.

The quantitative changes in land use in Devon Ranch as elicited through GIS analysis are summarised in Table 12.2. In this particular village, there was no settlement prior to designation and the total woodland area (encompassing the four sub-categories) has increased over the period (which includes five years of the resettlement programme). In continuity with the broad experience shown in Table 12.1, the biggest land use change under the course of the programme, has been the loss of grassland area.

Table 12.2 The nature and extent of land use changes in Devon Ranch, Tokwe Resettlement Area, 1972-85

	Percentage Area in 1972	*Percentage Area in 1985*	*Percentage Change 1972-85*
Cultivation	0	13	+13
Grassland	52	33	-19
Total woody cover	48	53	+5

Source: Elliott, 1995

The dynamism of land use at the village level is confirmed in Figure 12.2. All the shaded areas indicate the locations of a change in land use status of some kind between the two time-points of the aerial photography. It is evident that few areas within the village remained under the same land use category at the two time-points. The most extensive changes appear to have occurred in the western portions and simple visual analysis seems to suggest some linearity to the changes (along a NW/SE axis).

Figure 12.3 displays solely the land use conversions in Devon Ranch which involved woodland (the four sub-categories again amalgamated in this illustration). Evidently, whilst the quantified change in woody cover as shown in Table 12.2, may be an increase of total woodland, some areas within the village experienced losses over this short time period.

Figure 12.2 Areas of land use change in Devon Ranch, 1972-85
Source: Elliott 1995

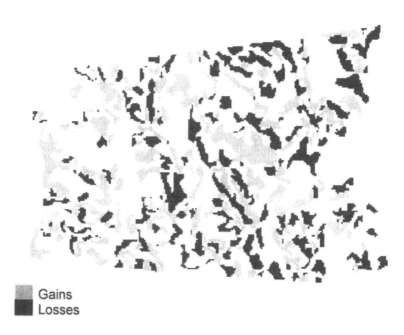

Gains
Losses

Figure 12.3 Woodland gains and losses in Devon Ranch, 1972-85
Source: Elliott 1995

Table 12.3 provides further insight to the patterns of woodland change and the associated land use conversions which have occurred. It is seen that 68 per cent of the woodland resources in 1972 remained in 1985. In contrast, only 43 per cent of grassland areas did not undergo a change in land use over the period. It can also be seen that, in the Devon Ranch case, cultivation (as a result of designation under the resettlement programme), occurred more widely in areas of former grassland, than they did in woodland areas, i.e. the opening up of areas for cultivation caused a greater loss of grazing area than woodland resources in this case. Furthermore, 28 per cent of the grassland areas in 1972 was reafforested by 1985, and indeed, exceeded the extent of land use conversions from woodland to grassland (21 per cent).

Table 12.3 The complexities of land use change (percentage of village area)

| | *Land use in 1985* | | | |
	Woodland	Grassland	Cultivation	Other
Land Use in 1972				
Woodland	68	21	6	5
Grassland	28	42	19	10

Source: Elliott, 1995

In summary, the GIS analysis of aerial photographs in the one case of Devon Ranch, has confirmed the complexity of the nature and direction of woodland changes over a time period including the implementation of land reform. In this case, there is no straightforward loss of woodland as a result of the expansion of cultivated area as suggested by Grundy *et al* (1993) in the Mutanda scheme. The significant differences found between the Devon Ranch case and in particular, the average situation for the 16 villages in Tokwe I and Wenimbi-Macheke generally, suggest that the pattern of woodland change with resettlement may be highly place-specific and that further case study analysis is essential. GIS analysis enables a rapid and potentially more accurate quantification of land use change with resettlement than offered through aerial photograph interpretation. However, it should be noted that substantial errors can occur both in the course of aerial photograph interpretation (on which the type of analysis presented here depends) and at the point of registration of aerial photography into a mapped format.

Processes of Woodland Change

For a full understanding of changes in miombo woodlands and in the livelihood systems derived therefrom, the changes have to be linked to processes at the local level (households, communities, stands of miombo woodland) and at the national, regional and global levels (macro-economic and water shed levels)'(Campbell *et al*, 1996).

At any point in time, local patterns of woodland use and management in resettlement areas will reflect a whole host of social and environmental factors, both within and beyond the land reform policy itself. For example, research in the communal sector has suggested strong differentiation in resource use even at the household level, including by variables such as gender, wealth and labour availability. With resettlement, not only are households exposed to different ecological contexts, but they are also subject to new institutional structures, including tenure conditions and local authority. All these factors at various scales could be expected to influence individual decision-making and resource use.

The complexity of the processes involved in household decisions regarding woodland use is illustrated in Figure 12.4. In a further element of the research already referred to (Elliott, 1994, 1995), interviewees reported the varied interplay of spiritual/religious factors, institutional controls, agricultural change and product demand, as well as woodland supply conditions, in their own explanations of woodland resource use (in this case with regard to brewing activities). It was evident that people could experience environmental change through movement to resettlement villages quite differently. Certainly, they ascribed varied importance to such perceived ecological change in shaping contemporary resource use decisions.

In summary, there is no simple model for explaining either how people experience ecological change with resettlement or how this affects the land use decisions which they make. There remains a substantial inter-disciplinary challenge to understand the varied land use practices in resettlement areas, how these change over time, and how (as noted by Campbell, 1996), these practices may or may not contribute to sustainable resource management.

- Here our crops are doing well so we need to boost our morale and also thank our gods for remembering us;

- How much wood we use depends on the type of beer we make;

- Usually in Chirumanzu we brewed beer to sell. Here we brew to thank God for our harvests;

- Because there is more wood here, we can brew for longer and make better tasting beer;

- I have not brewed beer since being resettled because I have not encountered any difficulties which necessitated brewing so as to consult our ancestors;

- We no longer brew beer because here there are no buyers;

- We don't brew any longer because we are not allowed to cut wood in the grazing areas;

- We no longer brew beer because we have discovered that beer brewing does not help us.

Figure 12.4 Perceptions of change in brewing activities on resettlement as reported by Tokwe I and Wenimbi-Macheke residents
Source: Elliott, 1995

Beyond the Boundaries

It has been suggested that a particular challenge of woodland resource management in the resettlement areas is to protect trees from claims of residents of adjacent communal areas (Bradley and McNamara, 1993). It should also be understood, however, that interactions across the tenurial boundaries occur in both directions and carry benefits as well as costs for the people and environments involved (Elliott, 1995). Kinsey (1986), for example, has documented the importance of interactions between settlers and their original communal area for the minimisation of the stress associated with their physical relocation. Retaining traditional land rights in communal areas and the pasturing of stock in scheme areas on behalf of relatives, were two principal practices found by Jinya (1991) in a study of resettlement permit violations. Unfortunately, much information, to date, regarding the extent of such 'resource-sharing' has been anecdotal

(Scoones and Matose, 1993). There is also a virtual absence of understanding concerning such relationships between resettlement and communal areas prior to land reform.

Research within the communal villages of Svosve and Shurugwi, neighbouring Wenimbi-Macheke and Tokwe I resettlement scheme areas respectively, found that visits for the purpose of woodfuel collection pre-dated the designation of those environments under the resettlement programme and had also increased subsequently (Elliott, 1995). This pattern was particularly strong in Svosve, where 47 per cent of households suggested that their main location for the collection of woodfuel currently was across the boundary in the Wenimbi-Macheke scheme area, as shown in Table 12.4. Ten years ago, only 25 per cent of the sample secured their needs in this way.

Table 12.4 The proportion of wood collected beyond the CA boundaries (per cent of respondents stating the RAs as the main source)

	Now	*Ten Years Ago*	*n=*
SHURUGWI	36	33	46
SVOSVE	47	25	50

Source: Elliott, 1995.

For the majority of respondents, however, access to these resources was perceived to have become more problematic since the conversion of land ownership under the resettlement programme. Residents of Svosve and Shurugwi CAs reported varied conditions on their access to resources in neighbouring RAs, as summarised below:

- Aggressive conflict with settlers;
- Collection only after dusk;
- Permission from Resettlement Officer;
- In exchange for labour on fields within RAs;
- Collection in designated areas only;
- No conditions;
- Through friends and family in RAs.

It was evident that despite the general increase in collection of woodfuel sources in the resettlement areas, the resource base for some communal residents had contracted in the last ten years, through their inability to access the required social channels (familial or 'official').

Implications for Policy

The presentation here of aspects of the research in the area of woodland resource use and management has given insight to the complexity of both pattern and process. The suggestion is clearly that environmental change through land reform is not simple, linear, nor uni-directional, as has been suggested in the past.

The continued significance of the woody biomass element in the livelihood systems of resettled households, confirms the central importance of integrating woodland resources into the future planning of land reform in Zimbabwe. Without such change, the prospects for meeting the immediate needs of settled households without resource degradation will be limited. Opportunities for sustained use of woodlands over the longer term are also likely to be compromised.

The evident place-specificity of process and pattern also confirms that future models for resettlement and the interpretation and implementation of the legislation, for example, need to be more flexible if the benefits of existing resource conditions and management practices are to be realised. This would include a recognition of and response to the real prospect that patterns of local ecology and resource use extend beyond the boundaries of administrative units.

References

Bradley, P.N. and McNamara, K. (eds.) (1993), *Living with Trees: Policies for Forestry Management in Zimbabwe*, World Bank Technical Paper No. 210, World Bank, Washington.

Campbell, B. (ed.) (1996), *The Miombo in transition: Woodlands and Welfare in Africa,* Center for International Forestry Research, Bogor, Indonesia.

Cliffe, L. (1988), 'The conservation issue in Zimbabwe', *Review of African Political Economy*, No. 42, pp.48-57.

Council of the Commercial Farmers' Union (1991), *The Land Issue*, paper presented to General Meeting of the Commercial Farmers' Union, Harare (11th January).

Elliott, J.A. (1994), 'The sustainability of household responses to fuelwood needs in the resettlement areas of Zimbabwe: a preliminary report of survey findings', Staffordshire University Occasional Paper, New Series A, *Geographical Research*, 3.

Elliott, J.A. (1995), 'Processes of interaction across Resettlement/Communal Area boundaries in Zimbabwe', *Geographical Journal of Zimbabwe*, 26, pp. 1-16.

Grundy, I.M., Campbell, B.M., Balebereho, S., Cunliffe, R., Tafangenyasha, C., Ferguson, R. and Parry, D. (1993), 'Availability and use of trees in Mutanda Resettlement Area, Zimbabwe', *Forest Ecology and Management*, 56, pp. 243-266.

Jinya, S.M. (1991), *Resettlement schemes permits administration and management: A diagnostic study of permit violations*, Resettlement Progress Report as at June, 30 1991, Ministry of Local Government, Rural and Urban Development, Harare.

Kinsey, B.H. (1986), *The socioeconomics of nutrition under stressful conditions: a study of resettlement and drought in Zimbabwe*, University of Zimbabwe Centre for Applied Social Sciences, Harare.

Moyo, S. (1995), *The Land Question in Zimbabwe*, SAPES Books, Harare, Zimbabwe

Scoones, I. & Matose, F. (1993), 'Local woodland management: constraints and opportunities for sustainable resource use', in P.N. Bradley and K. McNamara (eds.) *Living with Trees: Policies for Forestry Management in Zimbabwe*, World Bank Technical Paper No. 210, World Bank, Washington.

Whitlow, J.R. (1979), 'A scenario of changes in subsistence land use and its relevance to the tribal areas of Zimbabwe', *Zambezia*, 7, 2, pp. 171-189.

Whitlow, J.R. (1988), *Land degradation in Zimbabwe a geographical study*, Department of Natural Resources, University of Zimbabwe, Government Printer, Harare.

13 Land Inheritance Issues in Zimbabwe Today

DR SIMON COLDHAM

In late 1993 a Land Tenure Commission (LTC) was established, under the chairmanship of Professor Rukuni, to inquire into appropriate agricultural land tenure systems. Its terms of reference were extremely wide-ranging and included the requirement that it 'examine the present inheritance system in each of the sub-sectors of agriculture and, on the basis of the proposed system of land tenure, recommend appropriate inheritance procedures to be followed' (LTC, 1994: vol. 1 p.vi). In its report, which was submitted to the President in October 1994, the Commission was outspokenly critical of the rules and practices governing intestate succession to land, and it is the purpose of this paper to consider these criticisms and to assess the extent to which they have been addressed by recent reforms of the law of succession.

Former Law of Intestate Succession

It was, and remains, uncommon for Africans to make written wills, and where a deceased African had married according to customary law, the applicable law at that time provided that his/her land should devolve on his/her heir at customary law in his *individual* (rather than in his/her representative) capacity (Government of Zimbabwe, 1981) and that his/her movable property should devolve according to customary law (Government of Zimbabwe, 1997: AEA s.68[1]). Where, however, the deceased had married under the Marriage Act, the Supreme Court had made a controversial ruling in 1992[1] to the effect that his/her estate was governed by the general law, i.e. the Deceased Estates Succession Act, under which it would be shared between the deceased's surviving spouse and children. Broadly speaking, therefore, it was the deceased's form of marriage that determined the applicable law of succession, and in the great majority of cases the estates of Africans were governed by customary law rather than the general law.

The Commission's Views on Land Inheritance

The Commission was particularly critical of the customary heir principle which had apparently been designed to prevent subdivision and which had in fact often been interpreted to give the eldest son absolute rights over the land regardless either of his interest in occupying or farming the land or of the wishes or claims of other family members. In some instances an heir residing in the city had sold the land to a third party over his family's head[2] (Dengu-Zvogbo *et al*, 1994). This problem seems to have been especially acute in the small-scale commercial farming areas (SSCF areas), but increasingly prevalent in the communal areas (CAs). Nor did the Commission find it satisfactory that land should devolve in one way while property required for the effective use of the land (e.g.livestock and farm tools/machinery) should devolve in another (LTC, 1994: vol 2, pp.301-2) Finally, the Commission was concerned about the adverse effects on land use that the subdivision and fragmentation of holdings on death might have (LTC, 1994: vol 2, pp126-134). However, the incidence of these problems obviously varied from area to area, and it is appropriate to consider the four farming areas, i.e. the large-ccale commercial farming areas (LSCF areas), the SSCF areas, the resettlement areas (the RAs) and the CAs separately, as the Commission did.

Not surprisingly, the Commission found no succession problems in the LSCF Areas though it anticipated that problems similar to those occurring in the SSCF Areas would surface in the future as more polygamous families occupied farms there and there arose conflicting expectations from the wives and children of the deceased (LTC, 1994: vol 2, p.295).

In the SSCF Areas the Commission found serious succession problems, deriving in part from the principle that land (freehold/leasehold) should devolve on the customary heir. The Commission reported 'prolonged inheritance disputes' particularly within polygamous families (LTC, 1994: vol 2, p.296). The position of widows is more vulnerable in the SSCF areas than anywhere else (Dengu-Zvogbo, 1994). The Commission seems to have been ambivalent in its attitude to succession practices, at one time criticising the subdivision and fragmentation of holdings that is occurring and at another stressing the importance of protecting widows and children from the loss of their lands; at one time underlining the need to ensure that the best person is responsible for farming the land, pointing out the threat to productivity posed by polygamy and encouraging the making of wills, and at another attacking the way in which customary heirs may ignore family rights and proposing that the customary law of succession be replaced by the general law in these areas.

In the RAs it seems that it is the Resettlement Officer who determines who shall take over the permit on the death of the permit-holder. This is often the widow (LTC, 1994: vol 2, p.299); indeed women are more involved in decision-making in the RAs than in other areas (Dengu-Zvogbo, 1994). However, this has not prevented subdivision occurring in breach of permit, leading to conditions similar to those of the CAs. The Commission proposes that permits should be replaced by long leases including an option to purchase. The Commission sees this as the ideal system for promoting indigenous commercial agriculture (LTC, 1994: vol 1, p.69). As in the case of the SSCF Areas, the Commission recommends that the making of wills should be encouraged and that the general law of intestacy should be applied to the RAs.

It is in the CAs that land inheritance rights are likely to be most sharply contested, not least because, unlike the three areas already considered, pressure on the land is widespread and documents of title rare. The Commission found 'increasing discrimination against widows and other dependents' in inheritance matters (LTC, 1994: vol 1, pp.30-33). The Commission notes the move from the extended family to the nuclear family as the key landholding unit and while it argues that the 'traditional freehold' provides the family with security of tenure over arable and residential land, it recognises the conflicts that are increasingly arising over communal resources such as water and grazing. The subdivision and fragmentation of holdings are common, and with the rapid emergence of a land market, social stratification associated with differential access to resources is becoming widespread. Again the Commission is ambivalent about these trends. On the one hand, it laments the increasing vulnerability of widows and other dependants and recommends that the customary heir should hold the land on trust to give effect to customary entitlements and that a head of household should not be allowed to sell land without the family's consent. On the other hand, it points out that customary restrictions on the alienation of land create inefficiencies in land and labour markets and that customary inheritance law does not ensure that 'the potential occupants of agricultural land...are the appropriate persons to farm that land' (LTC, 1994: vol.2, p.293). While the Commission favours the registration of 'traditional freeholds' and the issuing of land registration certificates, it baulks at the idea of converting these rights into general law freeholds (LTC, 1994: vol.2, p.302ff).

While the Commission was successful in identifying the land inheritance issues that had arisen in each of the four areas, its proposals for addressing these issues lack conviction. For example, it recommends that people be encouraged to make wills in the SSCF areas and in the RAs, but it gives no reasons for this. It is true that increased use of the power to make wills might lead to a reduction in the number of inheritance disputes, but it would not necessarily resolve any other problems. Indeed, it is far from certain that the number of disputes would decline. When African men make wills, they do not usually leave their estates to their widows (Dengu-Zvogbo, 1994). Spouses and dependants may apply under the Deceased Persons Family Maintenance Act for provision to be made for them from the deceased's estate, but in practice applications are uncommon. The Commission endorses the retention of the dual system of intestate succession law (i.e. the general law and customary law) based on the deceased's form of marriage, though its proposals are neither supported by argument nor fully worked-out. It seems to favour a link between land tenure law and intestate succession law so that the general law of intestacy would apply in those areas governed by the general land law (i.e. the LSCF Areas, the SSCF Areas and, under its proposals, the RAs), while the customary law of intestacy would apply in the CAs. The applicable law of intestacy would depend not on the deceased's form of marriage, but on the area where the deceased had land rights. The Commission does not explain why the general law of intestacy would solve the problems arising in the SSCF Areas, nor does it seem to appreciate that an intestacy law applies not only to land but to all kinds of property.

New Law of Intestate Succession

While the Commission's recommendations on land tenure reform have yet to be implemented, its recommendations on land inheritance have been overtaken by the enactment of the Administration of Estates Amendment Act 1997 (AEAA: see Coldham, 1998) based, in its broad outlines, on proposals set out in the Government of Zimbabwe's White Paper on Marriage and Inheritance 1993. The dual system of intestacy law based on the form of marriage has been retained, so while the general law (based on the Deceased Estates Succession Act) continues to apply to the intestate estates of those who marry under the Marriage Act, the reforms will normally apply to the estates of those married according to customary law, i.e. the large majority of Zimbabwean Africans. The new Act makes no distinction between land (of whatever tenure) and other forms of property.

It will be interesting to see how far it will solve the land inheritance problems identified by the Commission.

One striking feature of the new law is that, rather than simply setting out the inheritance rights of various categories of beneficiary, it gives priority to consultation procedures and to the desirability of securing the agreement of the affected parties to the scheme of distribution. Thus the Master (usually a magistrate) is required to summon the deceased's family together for the purpose of appointing an executor of his/her estate, and the executor is required to draw up an inheritance plan providing for the conservation of the estate, for its distribution, for the sale of any property for the benefit of the beneficiaries and for the maintenance of any beneficiary (Government of Zimbabwe, 1997: AEA s.68D[1]). In drawing up the plan the executor must pay regard to the principles of distribution (below) and he must consult the deceased's family and the beneficiaries and endeavour to obtain the beneficiaries' agreement to the plan (Government of Zimbabwe, 1997: AEA s.68D[2]). Once the Master is satisfied that appropriate consultations have taken place and that the beneficiaries have agreed to the plan 'with full knowledge and understanding of their rights', he may then approve the plan (Government of Zimbabwe, 1997: AEA s.68E). Where, however, one or more beneficiaries do not agree with the plan, he must take steps to resolve the dispute and in so doing he must be guided by the principles of distribution set out in section 68F[2]. The Master may exempt small estates (currently defined as estates valued at less than Z$60,000) from the Act. They may then be administered and distributed informally (Government of Zimbabwe, 1997: AEA s.68H).

The principles of distribution are a little complex and not well drafted, but the following points should be noted. Any surviving spouse/s is/are entitled to the household goods and to ownership/usufruct over the house s/he was living at the deceased's death. (This is designed to protect the widow from eviction on her husband's death, and is in line with the Commission's own proposals.) However, the surviving spouse is not entitled to a minimum 'sum' from the estate; rather, the residue of the estate is to be divided between the surviving spouse/s and child/children in certain fixed proportions and regardless of gender/marital status/'legitimacy'. In the absence of a spouse or child it is divided between surviving parents, brothers and sisters in equal shares, and in the absence of these categories of relative it will devolve in accordance with customary law. While, therefore, customary law remains the residual law, it is clear that the effect of this reform is to abolish the customary law of succession and to replace it with a scheme that is based on gender equality and concentrates inheritance rights within the patrilineal nuclear family. Indeed, given that the principles of distribution do not diverge markedly from the general law position under Deceased Estates Succession Act, it is surprising that the government has retained the dual system based on the form of marriage and on a complex choice of law principles.

Except in those relatively uncommon cases where the deceased intestate had married under the Marriage Act, it is clear that land inheritance rights will now generally be governed by the new Act, whether the land is situated in the CA, in the RA, in the SSCF Areas or in the LSCF Areas, even though the land tenure systems operating in each of these four areas are markedly different. The Act addresses several of the problems raised in the Commission's report. The 'customary heir' principle is abolished, the customary claims of the extended family and of a deceased woman's natal family are largely extinguished in favour of the claims of the nuclear family, and the position of widows (particularly in relation to the matrimonial home), sons and daughters is strengthened. However, given that land, together with livestock and farm equipment, is likely to form the bulk of most estates, the question will arise whether the beneficiaries will be able to agree on an inheritance plan under the Act and, if not, how the Master will resolve an inheritance dispute. Even though members of the wider family (e.g. uncles, aunts and cousins) are not classified as beneficiaries under the Act, it seems certain that the further subdivision and/or the alienation (temporary or permanent) of at least part of the land will occur. The Commission points out that in the case of estates distributed under the Deceased Estates Succession Act, the principle that the surviving spouses and children share equally may lead to the sale and/or subdivision of the land in the absence of a 'distribution compromise'. (LTC, 1994: vol.2, p.290). The claims of those beneficiaries occupying the land will have to be reconciled with the competing claims of those who are not. Considerations of productive land use will have to be weighed against the traditional rights of access to land conferred by membership of the kin-group. Given the overriding desirability of securing agreement on distribution and given the principle that, 'so far as possible, the net estate should be applied to meeting the basic needs of beneficiaries who have no other means of support' (Government of Zimbabwe, 1994: AEA s.68F[2i]), agreements seem likely to uphold such rights of access.

One final point. Zimbabwe has a uniform law of testate succession and now, with the enactment of the AEAA 1997 and its replacement of customary law with a code that shares many features of the general law, it has something close to a uniform law of intestate succession. Would it not be consistent with this approach to abolish customary land tenure in the CAs and replace it with the freehold/leasehold title system that operates in the LSCF and SSCF Areas and that the Commission recommends to be introduced in the RAs, thus creating a uniform system of land tenure? Now that the needs of a deceased's surviving spouse/s and child/ren are addressed by the AEAA, a formal record of rights should be set up to afford them some protection during his/her lifetime and indeed to protect families from having their land acquired or disposed of by state or village institutions. The precise nature of the title introduced (freehold or leasehold) and the question whether controls should be imposed on the use or disposition of land are controversial issues that lie outside the scope of this paper.

Notes

1   Mujowa v Chogugudza, SC 142/92
2   See Seva v Dzudza, SC 131/92 and Mudzinganyama v Ndambakuwa, SC 50/93. In Masango v Masango, SC 66/86 it was held that the heir still owes a duty of support to the dependants of the deceased.

References

Coldham, S. (1998), 'Succession Law Reform in Zimbabwe', *Journal of African Law*, 42, 1.
Dengu-Zvogbo, K *et al* (1994), *Inheritance in Zimbabwe: Law, Customs and Practice*, Women and Law in Southern Africa Research Trust, Harare.
Government of Zimbabwe (1981), *Customary Law and Primary Courts Act*, no.6 of 1981, s.6 (A).
Government of Zimbabwe (1997), *Administration of Estates Act*.
Land Tenure Commission (1994), *Report of the Commission of Inquiry into Appropriate Agricultural Land Tenure Systems*, Government Printers, Harare.

14 Land Reform versus Customary Law: What About Women?

JENNY BROWN

The Issue

In rural areas of the south, land is of fundamental importance. It provides opportunities for shelter, subsistence and commercial activity and also for status, dignity and a sense of belonging and of self-worth. Often, these opportunities are denied to women except in their role of wife or mother.

Shona women in rural Zimbabwe are among those millions of African women who cannot in practice enjoy independent rights to land. Their communities treat them as perpetual minors, unable to own land or most other types of property (May, 1983). The implications of such incapacity can be shattering.

Since Independence in 1980, there has been a spate of legislation by the Government of Zimbabwe, apparently intended to give women a legal status equal to that of men. One hopes that the same intention will be seen in the current proposals for land reallocation, but the outcome at present can only be guessed at.

The 'Cultural Lag'

Zimbabwe, like many African states, has a pluralist legal system, comprising the general law (legislation, for present purposes) and 'customary law'. This latter phrase is commonly used to describe the regulatory and normative basis, often independent of legislation, on which large parts of the population conduct their lives.

Although law is often referred to as a potential tool for 'social engineering' (May, 1983) its transformational effect on society is not automatic and it has been indicated that for change to happen there must exist not only an awareness and understanding of the law, but also a willingness to use it (Cutshall, 1991). Where discriminatory attitudes persist, awareness and understanding of the law will achieve little unless members of the 'target' population choose to implement it. They will not however, if they see no benefit for themselves - so men may not voluntarily comply with the government's gender equality policies by, for instance, allowing their widowed daughters to farm family land. Nor will women exercise their legal rights if they are intimidated by procedures, jargon or personnel, or by other members of their community, nor if they are economically or geographically unable to access the enforcement procedures (Banda, 1992).

The concept of 'lag' was described by Dror, (quoted in May, 1983), it '...applies to law and social change in dynamic situations, after either social change or changes in the law occur and no parallel changes and adjustment processes take place in the law and society respectively.' Much Zimbabwean legislation purporting to give women equal rights appears neither to reflect prior changes in the gender ideology of the rural population nor to have been subsequently reflected in social changes. Men still consider women to be inferior and incapable of independent action and women often believe themselves to be so ('A man is more enterprising...I am just a woman' - Mrs Tsara, quoted in WLSA, 1994). Hence, a 'lag'.

What is it that needs to be changed in order to achieve greater gender equality in reality? Customary law has always been capable of change (Chanock, 1985). Fallers (1969), for example, talks about the 'efflorescence' of customary law, specifically in the context of land, as the socio-economic context changed during the colonial period. Challenges are increasingly being made by women to the patriarchal values which keep them subordinated, both in customary law and through legislation. The 'cultural lag' shows, however, little sign of diminishing.

Ultimately, the achievement of any state's legislative purpose is dependent on its acceptance into the communities which it is intended to affect. Changes may therefore be required to the customary laws which regulate life in those communities. This is clearly the case in the context of the Zimbabwean Land Reform Programme. One key question is whether legislation should reflect existing social values, implying delay until appropriate change has happened, or whether it should anticipate and try to influence such change. In the latter case 'lag' is inevitable. One can speculate as to whether the Government of Zimbabwe is in fact working towards a change in customary law. Having said that, the significance of legislation should not be underrated as itself being an agent of social change. As Hutchins comments (May, 1983) law '...is the way in which newly discovered moral truth is disseminated...law helps make the mores...'.

Much of the existing legislation is apparently radical in gender terms. The most potentially far-reaching enactment is The Legal Age of Majority Act 1982 (LAMA), which gave full legal competence to all Zimbabweans at the age of 18, irrespective of sex. Despite government denials this is fundamentally at variance with traditional Shona gender ideology. 'Widespread indignation' resulted from the realisation of the potential scope of LAMA following the decision in Katekwe -v- Muchabaiwa (1984) (Bennett, 1995), and judicial decisions on LAMA since 1984 have varied considerably in the extent to which they apply its principles in a family context, reflecting perhaps an ambiguous attitude on the part of the Government of Zimbabwe.

Zimbabwe has ratified all the major international human rights instruments. Article 2 of the key document on gender discrimination, the 1979 Convention on the Elimination of All Forms of Discrimination Against Women, provides that states should take 'appropriate measures' to eliminate discrimination and that these include, but extend beyond, legislation. The necessary abolition of discriminatory customs and practices (Article 2[f]) appears not yet to have been seriously tackled.

A gap is thus evident between the theory, namely the legislated equality of rights for all Zimbabweans over 18, and the practice, in which a patriarchal Shona society, governed by customary law, excludes women from independent rights to land and maintains them in a state of perpetual legal incapacity (May, 1983). This does not bode well for the land reform currently under way.

The Historical Context

Until recently the tendency among writers was to assume a pre-colonial 'Merrie Africa' (Chanock, 1985), but it is now accepted that pre-colonial societies had some form of social regulation, this being regarded as necessary for survival (Roberts, 1979).

It is difficult to ascertain precisely the status of pre-colonial women, given in particular the absence of an indigenous literate tradition (May, 1983). Whether society was then regulated by formal 'law' or informal norms and values, the position of women, including with respect to land, was very different from that during and after colonial times. Provided land remained abundant, which it generally did (Bennett, 1995), women's position was 'not necessarily disadvantageous' (Batezat & Mwalo, 1989). In practice the local chief would allocate land to the male head of household who would then allocate parcels to each of his wives. They would use it for subsistence purposes, but would be allowed to keep any surplus. This relative freedom is confirmed by May (1983), but should not obscure the fact that women remained political and jural minors throughout their lives.

Colonialism made many fundamental changes. The authorities, being Victorian middle-class men, made certain erroneous assumptions about the indigenous societies; (Chanock, 1982); these included that women were, as in England, institutionally subordinated and that it was accordingly appropriate to deal exclusively with African men in determining the 'customary law' which was to form the basis for social regulation. 'Male claims...were legitimated as "customary"' and women's claims, not being 'traditional', could not be effective. A new 'customary law' was created and imposed by reference to an idealised past (Snyder, 1981), one which in fact bore little resemblance to actual practices.

Bates (1995) comments that since rights in property limit other people's rights and ability to use that property, 'to alter property rights is to redefine social relationships.' This is what appears to have happened in the Shona communities; customary law was constructed for political and administrative purposes by the (male) British authorities, with the willing assistance of African men, to their mutual benefit. Land rights were transformed by this process and by that of commercialisation, excluding women even from such informal rights as they had previously enjoyed. Women's inferiority was institutionalised during the colonial era, and reinforced by the superior access to formal and legal rights which men received (May, 1983).

In the early years of independence ZANU-PF was keen to be seen to be pursuing socialist policies and to be reforming customary practices where these were inconsistent with those policies (Ncube, 1991). There appears, however, to have been a conflict between two incompatible attitudes - the desire to recognise women's contribution to the liberation struggle on the one hand and, on the other, a 'much romanticised' view of customary law (Banda, 1992) which ultimately seems to have prevailed. Now, the government's gender policies seem far less radical and it seems to be bending to conservative male pressure, as in relation to the proposed land tenure reforms in respect of which it appears to have lost its nerve when faced with the opportunity to transform women's position. Complaints have been made, for instance, in the context of the report of the Land Tenure Commission, that women were not fully consulted during the Commission's deliberations (WLSA, 1996).

The transformational potential of the legislative programme is not to be discounted in the quest for greater gender equality as far as land rights are concerned. But something more is required if women are fully to achieve equality of status in practice as well as in theory. Is the Government of Zimbabwe prepared to follow through on this legislation and take steps to ensure that the principles of gender equality apparently enshrined in it are assimilated at grass-roots level?

Prospects and Challenges

The need perceived by many Zimbabwean women is for independent access to land. The government has undertaken extensive commitments to gender equality in its legislative programme. These commitments appear to have been implemented, if at all, with little success; customary law continues to operate largely independently of the state, and 'off the record' practices persist (An-Na'im, 1994). The 'stickiness' of customary law remains possibly the biggest challenge for any government aiming for effective transformation of gender relations.

How can the range of choices available to women be increased? Many women will, given the opportunity, choose to exercise independent land rights, and many are increasingly seeking change (Zimrights, 1996a, 1996b). Some will not - perhaps for strategic reasons, such as to avoid falling out with male relatives who might help them in the future (WLSA, 1995). The choice should, however, be an individual and informed one. It should also be genuine, 'not manipulated or limited by discriminatory legislation or practices - or by the lack of social, economic or cultural alternatives' (Holm, 1995). Too often relevant information is lacking, for both men and women, and if this problem could be dealt with much progress towards gender equality could be achieved.

The Government of Zimbabwe can make a major contribution to the desired changes of customary law institutions. This could be facilitated if more women became involved in the legislative and government systems as lawyers, politicians or civil servants, but this will not be enough- and there is always the possibility that such women may not adequately represent or understand the interests of rural, poorly-educated women (WLSA, 1995). More girls should receive at least a basic formal education. More should be done to render the state judicial system accessible and user-friendly, particularly for women, to counter the prevalent economic, geographic and psychological barriers to the exercise of their rights (Banda, 1992). In the specific context of land reform, awareness programmes will be required so that all sections of the population are fully informed about the schemes and opportunities. State employees administering the reallocation should be trained to ensure that women in practice have equal opportunities, and are positively encouraged, to apply for their own land, and that following allocation women receive all appropriate training, inputs and support (bearing in mind that women's needs may vary from those of men).

Necessary as such measures are, in terms of the signals that they would send about the government's commitment to gender equality, and in terms of the formal framework which they would create, they are not on their own sufficient to generate fundamental changes at grass-roots level. It is important also to identify actions which can be taken outside government to complement and support its own activities.

Education is key. In their legal education programmes for women WLSA (1995) found 'an insatiable appetite for knowledge', and they agree that legislation is only one factor in the gendering process. The aim should be to encourage women to question and challenge the status quo.

Such encouragement forms part of a wider process of empowerment. In this context the term implies the development of a positive self-perception (WLSA, 1994) and sense of self-worth and identity (Elson, 1991). Collective action is in practice key to the empowerment process. Mackenzie (1990) describes how Kenyan women responded to male control over land and labour, justified by kinship ideology, by adapting a customary practice of sex-specific working which enabled women's groups to purchase land.

If change is to be achieved, education of men is also vital. To be faced with a community of newly-empowered women is not an attractive prospect and could result in even further polarisation of attitudes as men feel a threat to their positions of power. Education is therefore necessary, to persuade men that loss of 'traditional' power can be compensated for by, for instance, increased efficiency and other benefits for the family unit (WLSA, 1995), including improved nutrition and better children's health, and more sustainable agricultural practices. In support of this approach Holm (1995) comments that unless women feel they have a meaningful stake in the land they farm 'they are unlikely to see any point in conservation schemes [or] tree-planting...Why should a woman make an investment...in an asset that may be lost to her at a moments notice?' Men should be encouraged to see the force of this type of argument.

The impetus for such learning will have to come from women - there is no obvious incentive for men to educate themselves out of power, as they would see it. Women will need self-confidence, and the information which gives that self-confidence. This will take courage and commitment on the part of the government, but if that support is forthcoming the results will be very exciting. If it is not, land reform will be yet another missed opportunity.

References

An-Na'im, A.A. (1994), 'State Responsibility to Change Religious and Customary Laws', in R.J. Cook (ed.) *Human Rights for Women: National and International Perspectives*, University of Pennsylvania Press, Pennsylvania.

Banda, F. (1992), *Women and Law in Zimbabwe: Access to Justice on Divorce*. PhD Thesis, University of Oxford, Oxford.

Bates, R.H. (1995), Social Dilemmas and Rational Individuals: An assessment of the new institutionalism, in Harriss, J. *et al* (eds), *The New Institutional Economics and Third World Development*, Routledge, London.

Batezat, E. and Mwalo, M. (1989), *Women in Zimbabwe*, SAPES Trust Harare.

Bennett, T.W. (1995*), Human Rights and African Customary Law Under the South African Constitution*, Juta & Co. Limited, Cape Town.

Chanock, M. (1982), 'Making Customary Law: Men, Women and Courts in Colonial Northern Rhodesia', in M.J. Hay and M. Wright (eds), *African Women & the Law: Historical Perspectives*, Boston University, Boston.

Chanock, M. (1985*), Law, Custom and Social Order: The Colonial Experience in Malawi and Zambia*, Cambridge University Press, Cambridge.

Cutshall, C.R. (1991*), Justice for the People: Community Courts and Legal Transformation in Zimbabwe*, University of Zimbabwe, Harare.

Elson, D. (ed.) (1991), *Male Bias in the Development Process*, Manchester University Press, Manchester.

Fallers, L.A. (1969), *Law Without Precedent: Legal Ideas in Action in the Courts of Colonial Busaga*, University of Chicago Press, Chicago and London.

Harriss, J., Hunter, J. and Lewis, C. (eds) (1995), *The New Institutional Economics and Third World Development*, Routledge.

Hay, M.J. and Wright, M. (eds) (1982), *African Women & the Law: Historical Perspectives*. Boston University, Boston.

Holms, G. (1995*), Women and Law in Southern Africa*, Ministry of Foreign Affairs, DANIDA, Copenhagen.

May, J. (1983), *Zimbabwean Women in Customary and Colonial Law*, Holmes McDougall Limited.

Mackenzie, F. (1990), 'Gender and Land Rights in Murang'a District', Kenya *in Journal of Peasant Studies* 17, 4.

Ncube, W. (1991), 'Dealing With Inequities in Customary Law: Action, Reaction and Social Change in Zimbabwe' in *International Journal of Law and the Family*.

Roberts, S.A. (1979), *Order and Dispute: an Introduction to Legal Anthropology*, Penguin Books Limited, Harmondsworth, Middlesex.

Snyder, F.G. (1981), 'Colonialism and Legal Form: The Creation of "Customary Law" in Senegal', *Journal of Legal Pluralism, 19*.

Women and Law in Southern Africa (1994), *Inheritance in Zimbabwe: Law, Customs and Practices,* SAPES Trust, Harare.

Women and Law in Southern Africa (1995a), *Widowhood, Inheritance Laws, Customs and Practices in Southern Africa.* Women and Law in Southern Africa Research & Educational Trust, Harare.

Women and Law in Southern Africa (1995b), *Beyond Research: WLSA In Action*, Women and Law in Southern Africa Research & Educational Trust, Harare.

Women and Law in Southern Africa (1996), *Regional Impact Study Report*, Women and Law in Southern Africa Research & Educational Trust, Harare.

Zimrights Women and Land Programme (1996a), *Progress Report Covering the Period June-August 1996,* Zimrights, Harare.

Zimrights Women and Land Rights Project (1996b), *Voices From the Fieldwork*, Zimrights, Harare.

15 The Effects of Land Reform on Gender Relations in Zimbabwe

DR SUSIE JACOBS

Gender matters are frequently 'written-out' of accounts of land reform, both historically and internationally. Likewise, gender issues within Zimbabwean land reform remain largely marginalised in policy terms. This is a great anomaly. Gender relations have profound effects in many spheres, including agricultural production and future prospects for land reform. Despite recent changes, and despite the fact that rural African women are voicing their needs more overtly, they remain disadvantaged in legal, economic and social respects.

This chapter examines the effects of resettlement upon gender relations and upon married and widowed women's lives. The findings of the article are based on research carried out in the 1980s in several resettlement areas (RAs) in north-eastern Zimbabwe. The paper argues that:

- Resettlement policies, although beginning to change by the 1990s, still operated to keep wives the dependents of husbands by assigning land to household 'heads';
- Widows, although at times marginalised, are assuming more important roles and are becoming more numerous;
- Not only gender but other factors such as class, age and marital status (among others) need to be examined in any evaluation of the position of 'women';
- Despite women's situation of *de facto* legal, economic and structural dependence, resettlement has nevertheless benefited wives and widows in several respects.

Research Methodologies

In 1983/84, research was undertaken on gender and the land resettlement programme. The research was attached to the (then) Ministry of Community

Development and Women's Affairs. In this paper, only aspects of the research relating to Model A resettlement schemes are discussed. The research employed several methodologies: a structured questionnaire; semi-structured interviews; observations; key informant interviews; and 24 group meetings with Women's Clubs in six RAs. Overall, 650-700 women were contacted in the meetings. Structured interviews, which form the basis for some of the more detailed data below, were conducted in eight villages of Mt. Darwin and Hoyuyu RAs in north-eastern Zimbabwe. The survey was multiple-stage and included 207 settlers, consisting of 99 married women, 66 men married to women in the sample, and 42 widows/divorcees. This was the first study of gender relations within resettlement large enough to employ statistical data.

Backgrounds

Settlers in the sample came from a variety of backgrounds and regions, two-thirds being from nearby communal areas (CAs). The remaining third were from distant CAs, commercial farms, from 'keeps', small-scale commercial farms, towns, or were refugees from abroad (mainly Mozambique). Settlers also had a variety of religious affiliation: nearly one third (32 per cent) followed traditional African religions; 22 per cent each were Roman Catholic and Apostolic (the latter being an African Christian church); and others belonged to Methodist, Anglican or Protestant fundamentalist or evangelical churches. Thus settlers underwent the experience of resettlement, itself often a dramatic change, and found themselves living with people from a variety of regional and religious backgrounds. An additional change, and one which was widely disliked, was the new pattern of settlement in villages rather than *kraals*; many settlers complained of 'crowdedness' as one of the worst aspects of resettlement.

Social Class Differentiation as a Gender Matter

One of the most important social divisions is that of class. Since Model A resettlement seeks to encourage the development of a stratum of self-sufficient or well-to-do peasants, it would be unusual were class differentiation not to develop. That is, people come to resettlement with varying and unequal resources, and social processes may accelerate the development of such inequality. The differentiation referred to is of a limited nature, thus, distinctions referred to are between strata *within* the peasantry.

In Zimbabwean rural areas, most women are seen as having a subordinate status. However, this factor does not mean that class differences, like other social differences, are irrelevant to them. For wives class is mediated by a variety of gender factors, in particular the marriage contract. Thus, class matters for women, but not necessarily in the same way as for men.

Using two measures of class, differences were found between strata of poor, middle and wealthier settlers. Both employed the measure termed 'self-sufficient': self-sufficiency in means of cultivation was defined as possession of two adult cattle and one plough (see Box 15.1). Both variables were calculated individually, and were related to factors such as education, religion, crop harvests, amount of debt and amount of expenditure. The relations between these variables and class measures were valid at levels of statistical significance of 5 per cent or below. Such statistical significance held true for husbands, wives and widows. The fact that measures of class are related to a *variety* of social factors tends to indicate that these are not just artifacts but are of some sociological importance in the population. For example, one feature that stood out was the position of ex-commercial farm labourers, as the least educated and most impoverished settler grouping.

Class 1: the variable Class 1 was calculated by addition of scores awarded to the following component variables:
i)   does wage labour = 0/no wage labour = 1
ii) no hiring in of labour = 0/hiring in of labour = 1
iii) less than self-sufficient = 0/self-sufficient in means of
     cultivation = 1
iv) cultivates 0-12 acres (i.e. amount allocated) = 0/cultivates 13+
     acres = 1

Class 2: the variable Class 2 was calculated by the addition of a score for a fifth variable to Class 1, this concerning receipt of wage remittances:
v)   Z$0 - 99 p.a. received in wage remittances = 0/receipt of
     Z$100+ p.a. = 1.

Box 15.1 Measures of social class used in the study

The stratum of 'poor peasants' consisted of peasants who were less than self-sufficient in their ability to cultivate. Despite having access to land since resettlement, they might lack sufficient draught animals or implements, and many did wage labour for others, regularly or seasonally. Many of these people were ex-commercial farm labourers. The 'wealthiest' stratum, on the other hand, had sufficient cattle and implements to plough; most hired in labour and one-fifth also managed to gain access to more than 13 acres of land, due to (illegal) sharecropping arrangements. Many received wage remittances, usually from children, and these were important to household and individual income levels. The middle stratum was the largest, constituting nearly 50 per cent of the sample population in these north-eastern RAs. A small majority were stereotypical 'middle peasants' in that they were independent of commodity production (they neither performed nor hired in wage labour nor received remittances). However, the other half of the middle stratum were involved in complex sets of commodity relationships, e.g. at once hiring in wage labour and performing wage labour themselves.

Measurement of class is never a straightforward matter, particularly as social class involves households as well as individuals within them. However, it is possible to make these points since an individual measure of class was used which could be disaggregated by sex.

One example of the importance of seeing social class as a gendered phenomenon, may be seen in wives' influence within households. An indicator, or index, of 'power' (see Box 15.2) or influence was constructed, using answers to 27 questions from the survey. The questions concerned matters such as wives' ability to make decisions or to influence them; the gender division of labour; behaviour and attitudes; and wives' perceptions of their own leverage within the home. Although less valid than long-term observations, this index provided a useful, if simplified measure. Using this measure middle peasant women's scores (i.e. their perceptions of leverage/authority, as measured) were much lower than those of wealthier and poorer peasant wives. The highest scores were those of wealthier women. The measured differences among strata were statistically significant. The types of assets owned or held by wealthier stratum women included shares of crops, livestock and chickens, and garden plots. (This argument is made by resource theory: e.g. Blood and Wolfe 1960; Safilios-Rothschild, 1970.) Why should such a pattern exist?

'Wealthier' peasant husbands can generally accumulate some capital, and usually hire in some labour rather than relying heavily on that of the wife. Wealthier stratum women in the sample also had more material assets themselves, and these resources may have contributed directly to their household influence. Poorer husbands lack resources with which to accumulate capital, and 41 per cent of poorer wives in the study did wage labour. It is possible that bringing in some income gave these wives more influence within their marriages. Additionally, there exists some evidence internationally that patterns of budgeting are most egalitarian in poorer households (e.g. Brannen and Wilson, 1987; Roldan, 1988).

A statistically significant relationship existed between Class1 and the index 'wifepower' (see text p.178); the latter was divided into 'low' and 'high' for purposes of cross-tabulation. At the 5% level of significance, with 2 degrees of freedom, the critical value of X^2 is 5.99. The observed value of X^2 was 6.11, indicating that a statistically significant relationship exists between the two variables.

Class2 could be computed for only 77 married women. At the 5% level of significance, with 2 degrees of freedom, the critical value of X^2 is 5.99; at the 2.5 % level of significance, with 2 degrees of freedom, the critical value of X^2 is 7.38. The observed value was 6.27, indicating that a significant relationship exists between Class2 and 'wifepower' for married women.

A statistically significant relationship also existed between 'husband's Class2' and 'wifepower' for married women. At the 2.5 level of significance, with 2 degrees of freedom, the critical value of X^2 is 7.38. The observed value was 7.93, indicating that a significant relationship exists between he two variables.

Additionally, when the indices 'class' and 'wifepower' were each calculated in a somewhat different manner and were cross-tabulated, the resulting statistical relationships always contained the same downward curve or dip, with middle stratum women scoring low compared with poor and wealthier women.

Box 15.2 Relationship between Class and 'Wifepower': statistical significance

Middle peasant men have only limited possibilities for accumulating resources, and must, in general, rely upon household labour power in order to effect this. Therefore middle peasant wives' labour is likely to be crucial to the farm. This stratum of women are likely to come under more direct supervision and to have their labour 'squeezed' hardest (see Box 15.2). This is a disturbing finding, given that the aim of resettlement is to encourage a 'self-sufficient' peasantry: such peasantries are often heavily exploitative of female labour.

Another point which can be drawn from these findings, is that 'women' should not be seen as an undifferentiated category; benefits of resettlement appear to accrue more to some groupings of women than to others.

Widows

One such difference was that of marital status. The position both of widows and of married women was examined. At the time this study was conducted, relatively few widows and fewer divorcees were resettled as female household heads. However, today the number of widows is increasing. This is due to the fact that within RAs, aspects of customary law are deemed not to apply, so that widows are now inheriting husbands' permits rather than these passing to the man's lineage relatives. Many of these widows are far younger than their husbands were, so that substantial numbers will inherit. Gaidzanwa (1995) writes that 15 per cent of permit holders are female; in Goebel's Wedza study (1999), 19 per cent of tenants were widows. However, Goebel points out that widows have no automatic rights to inherit. Such decisions rest with the Resettlement Officer, hence are variable, further implying that entitlements are unstable.

The increase in the number of widows, however, means that any data concerning widows may be of relevance in the future. In this sample, widows differed in their religion from married people, in that over half belonged to 'established Christian' churches (e.g. Roman Catholic, Methodist). Studies of CAs have consistently indicated that widows are, in general, an impoverished grouping. This was also the case for the majority of those interviewed here, in that 80 per cent had, before resettlement, earned between Z$0.00 - Z$100, with half of these having had no income. However, the remaining 20 per cent contained a few women who had earned up to Z$1,000 previous to resettlement. Post-settlement, 12 per cent of widows earned that amount or more. Thus, despite the general social difficulties widows face (see below), some are economically privileged. Nevertheless, the great majority remain poorer than other settlers.

In this survey, the types of concerns expressed by widows were similar to those mentioned by male heads of household: nearly all mentioned dirty and distant water supplies, poor transport, lack of shops, schools and markets, as well as lack of opportunities for wage labour. However, widows face serious problems of recruiting and commanding labour. As might be expected, they also commonly suffer from feelings of isolation. In informal contexts, for instance, many claimed that married women in the new villages suspected them of trying to steal their husbands. However, a perhaps more surprising finding of this study is that, at least at this stage of resettlement, widows suffered from such feelings little more than married women. In general, few settlers, especially female ones, found the atmosphere in their villages supportive or congenial, since informal networks of support were not well-established. Widows felt themselves, however, to be more dependent on social networks than did wives.

Married Women and Resettlement: Positive Aspects

The majority of female settlers are married women. The policy of assigning land to 'household heads', normally men, remains the most important determinant of married women's position, since upon divorce they lose rights to stay within RAs. In this way official policies view households as undifferentiated entities and perpetuate wives' position of insecurity. The suggestion that land reform may disadvantage many married women is in line with international experience (Jacobs, 1997; Palmer, 1985). Despite this, there exist features of resettlement which can be seen as positive for married women, and which mark Zimbabwe out as an unusual case in the history of land reform.

Gender Division of Labour

The first positive aspect concerns a shift in the gender division of labour. Even though most settlers worked harder than previously, due mainly to having more land, settler men reportedly did far more fieldwork than had been the case in CAs. In particular, men assist more with ploughing as well as with sowing, weeding and harvesting. The most common situation did not seem to be that husbands took over tasks customarily considered 'male', but which have become feminised in many CAs, but that husbands and wives worked together in fieldwork. A number of men also reported assisting with 'female' tasks such as fetching water and firewood. A very small proportion of men assisted with housework; this is important, given the stigma attaching to male participation in domestic labour. Wives commonly said, 'We work together here as a unity'. However, this phrase should not be taken too literally, given that (most) women's overall work burden, involving as it does nearly all domestic labour and childcare as well as longer hours of work in agriculture, remains much higher than (most) men's. Nevertheless, the burden of agricultural labour has been lightened for many wives, who feel this to be a highly positive development.

Property and Income

Wives in the sample had improved situations regarding property and income compared with their holdings previous to resettlement. Firstly, 37 per cent of wives had been allotted land of 'their own' to cultivate. Many of these had not held land previously. Amounts of land held varied from 0.5 to 4 acres, with 1.6 being the mean. More recent work in RAs has show much higher percentages of wives being allocated land, one study citing nearly all wives as having land (Chimedza, 1988); Goebel (1999) found that 65 per cent of wives had their 'own' plots. Such differences may simply relate to variations between sample populations or may indicate improvement over time.

A minority of wives in the sample were allocated harvested crops for use or sale by husbands. Additionally, about one-quarter of men and wives said that wives had been allowed to use or keep proceeds from specific crops such as cotton or groundnuts.

Wives own income in this population appeared to be relatively high, averaging Z$173.00. However, 20 per cent of wives had no income. The most important reported income sources consisted of sales from maize and cotton (50 per cent of wives' income); income from handicrafts, beer-brewing, herbalism etc. (10 per cent), and sales of eggs (5 per cent). Wives' average income amounted to one-quarter of mean-reported household income, a high figure which may represent a surprisingly (and welcome) high degree of intra-household redistribution. Kinsey's work (1998; Chapter 9 of this text) also indicates that settlers have prospered in relative terms over time. Thus, despite differences of class, and despite the fact that the extent of redistribution varies between households, it is likely that the RA wives surveyed are better off materially than are many CA women.

Family Change, Good Husbands and Resettlement Officers

Changes in family structure, organisation and processes, generally so central to women's lives, were significant to the generally positive feelings towards resettlement expressed by many wives. Related to these changes, many wives viewed their husbands as behaving in a more satisfactory manner in RAs than they had previously.

In general, the resettled people in my sample did not live in three-generation households. Although about half of households included dependent relatives, most of these were siblings of adult settlers or relatives' children. 12 per cent of households contained relatives of the senior generation. The majority of settlers, both male and female, expressed relief at living away from extended families. 70 per cent of husbands and 60 per cent of wives said that the nuclear family was the 'best unit in which to live', and many people said that extended families had caused great friction in their lives. The main reasons given for dislike of extended families were the drain on resources they can entail, and, more importantly, the interference of the older generation in matters now perceived as individual, or pertaining only to nuclear families. For many wives, resettlement meant an increase in influence. In the absence of other relatives, and 'living among strangers', husbands may draw closer to their wives and may adopt a more companionate model of family life.

This is one possible reaction, and it appeared to be a common one. However, another development which should be noted is the increase in polygynous households within RAs. I found rates of over 30 per cent, and Chenaux-Repond (1994) found a rate of up to 36 per cent. This phenomenon tends to indicate that many junior wives in RAs are being treated as labourers, similar to the phenomenon found by Weinrich (1975) and by Cheater (1981) in small-scale commercial farms. It is unlikely that these wives benefit from any developments towards companionate marriage.

The majority of wives' views, as expressed in the questionnaire administered, in less structured interviews and in Women's Club meetings, was that their male partners were now 'better husbands': that is, they drank less; worked harder; directed more resources towards the household/farm (rather than to other women); were less violent and esteemed wives more. The model of a more self-contained nuclear family with a 'good' husband, it should be noted, does not imply female autonomy or equality. As Bell and Newby (1976) wrote, '...the history of ideas about the Good Husband would reveal continual change and yet this has not threatened the traditional authority of husband over wife'. Again, differences of class were important. Middle peasant women were more likely than 'poor' or 'wealthier' women to say that their household status had declined or had remained unaltered post-resettlement and that husbands took decisions over cropping, budgeting, etc. without consulting them at all.

For the majority who perceived that their status had improved, several lines of explanation may be relevant. One set of factors is contingent: beer halls are usually much further away and opportunities to carry out extra-marital relationships, perhaps less. As noted, the predominance of the nuclear family may have improved wives' position, at least among monogamous families. Perhaps most importantly, both men and women know that their behaviour is scrutinised by Resettlement Officers (ROs) who potentially hold a great deal of power. Male title-holders know that they stand to lose permits if not, as it was said, 'on good behaviour'. In practice, among the ROs I interviewed, settlers are only removed for gross misbehaviour, but the possibility of removal appeared to be felt as a deterrent. Additionally, some ROs imparted a model of 'proper' family behaviour: in some cases this consists simply of hard work, investment in the farm, and not disrupting neighbours. For others, a notion of the wife as companion and as due respect, was an element. Even the former, more limited view, of proper behaviour tends to encourage men to be sober husbands as well as good farmers. In this instance, social and state control of a direct kind, as well as familistic ideology of a more indirect nature, may have operated to the benefit of married women.

Conclusion

Resettlement has altered and improved women's lives in various ways. Many wives and widows in this study gained in income terms. More recently, Resettlement Officers have often used their discretion to rule that permits be inherited by the widow, not by husbands' patrilineal relatives (refer to Chapter 14 of this text for further discussion of this point). At the same time, most structural components of subordination remain, so that wives are still viewed as dependents of husbands. Within this framework, their experiences are not uniform. This study suggests that one source of social difference is class: middle stratum rural wives' household status may be particularly negatively affected by the increased dependence on their labour.

Without firmer rights to land, both wives and widows will remain dependent upon the goodwill of Resettlement Officers and of husbands. Rural women are increasingly demanding such rights (Chenaux-Repond, 1995). Zimbabwean resettlement already stands out as a generally positive case internationally. If Zimbabwe were to grant married and unmarried women land access and rights which do not have to be mediated through male kin, this would set an example in the history of gender and land reform. It is hoped that future state policy will recognise women's crucial roles as farmers, and the importance of secure land access for their lives and livelihoods. However, any rights women have gained in land reform programs in other international contexts have only come about as part of movements for greater democracy, accountability and equality (Jacobs, 1998).

References

Bell, C. and Newby, H. (1976), 'Husbands and Wives: Dynamics of the Deferential Dialectic' in Barker, D. and Allen, S. (eds.), *Dependence and Exploitation in Work and Marriage*, Longmans Publishing, Harlow.

Blood, R. and Wolfe, D. (1960), *Husbands and Wives*, Free Press, New York.

Brannen, J. and Wilson, G. (eds) (1987), *Give and Take in Families*, Allen and Unwin, London.

Cheater, A. (1981), 'Women and their Participation in Commercial Agricultural Production', *Development and Change*, 12 (July).

Chenaux-Repond, M. (1994), 'Gender-Based Land Use-Rights in Model A Resettlement Schemes of Mashonaland, Zimbabwe', ZWCN Monographs, Harare.

Chenaux-Repond, M. (1995), 'Women Farmer's Position Paper', 3rd Draft, Harare.

Chimedza, R. (1988), 'Women's Access to and Control over Land: the Case of Zimbabwe', Dept. of Agricultural Economics *Working Paper* AEE 10/88, University of Zimbabwe.

Gaidzanwa, R. (1995), 'Land and the Economic Empowerment of Women: a Gendered Analysis' *SAFERE*, 1, 1.

Goebel, A. (1997), *No Spirits Control the Trees: History, Culture and Gender in the Social Forest in a Zimbabwean Resettlement Area* Ph.D. dissertation, Department of Sociology, University of Alberta.

Goebel, A. (1999), 'Here it is our Own Land, the Two of Us: Women, Men and Land in a Zimbabwean Resettlement Area' (forthcoming), *J. Contemporary African Studies*.

Jacobs, S. (1989) *Gender Divisions and Land Resettlement in Zimbabwe*, D. Phil thesis, Institute of Development Studies at the University of Sussex, Falmer, Brighton.

Jacobs, S. (1997), 'Land to the Tiller? Gender Relations and Types of Land Reform', *Society in Transition* (formerly, *J. of South African Sociology*), vol. 1, pp. 1-4.

Jacobs, S. (1998), 'The Gendered Politics of Land Reform: Three Case Studies', in V. Randall and G. Waylen (eds.), *Gender, Politics and the State*, Routledge Press, London.

Kinsey, B. (1998), *Allowing Land Reform to Work in Southern Africa: A Long Term Perspective on Rural Restructuring in Zimbabwe'*, Conference on Land Tenure in Developing Countries, with Special reference to southern Africa, Cape Town.

Palmer, I. (1985), *Women's Roles and Gender Difference in Development: The NEMOW Case*, Kumarian Press, New Haven, Connecticut.

Roldan, H. (1988), 'Renegotiating the Marital Contract: Intra-household Patterns of Money Allocation among Domestic Out-workers in Mexico City', in D. Dwyer, and J. Bruce (eds), *A Home Divided*, Stanford University Press, Stanford.

Safilios-Rothschild, C. (1970), 'The Study of Family Power Structure: a Review 1960-69', *J. of Marriage and the Family*, 31, 2.

Weinrich, A.K.H. (1975), *African Farmers in Rhodesia*, Oxford University Press, Oxford.

16 Land Reform in Zimbabwe: Dimensions of a Reformed Land Structure

PROFESSOR MANDIVAMBA RUKUNI

Introduction

Since 1890 up to today, the land question has singularly had the most significant impact on Zimbabwe's political and economic history. It can be argued, therefore, that a national solution to the land issue may well be the single most important national instrument as we enter the 21st Century. In addition, tenure security in terms of exclusive land rights of groups and/or individuals is the very basis of economic, political and social power and status. This is why it is agreed worldwide that land reform is essentially a political process. That as it may, land reform has to meet more than political objectives; land reform has to provide a solid basis for long term economic growth and social integration. To achieve this noble set of objectives, land resettlement or redistribution alone is no longer adequate, and it is more appropriate to plan and invest for a 'land and agrarian reform'. This basically means transforming land distribution patterns, strengthening security of land tenure, and strengthening rural institutions that manage land administration, and provide economic services to land users. This way, political, economic and social objectives can be met.

Following independence in 1980, the Land Resettlement Programme resulted in one of Africa's most successful examples of land redistribution. A total of 3.3 million hectares of land has been resettled to approximately 60,000 smallholder households. No other African country has acquired this amount of land from private landowners and re-distributed it to the poor and landless. In the short to medium-term, however, the Land Resettlement Programme of the 1980s only partially addressed the serious problem of land hunger, poverty and unemployment. A new strategy for land reform should therefore emphasise economic growth and development, and provide a solid foundation for rural economic empowerment. The land resettlement programme eventually slowed down by the mid-1980s and some of the reasons included cost of land, decreasing ability of government to finance the programme, inadequate institutional capacity, and inconclusive evidence of successful farming.

The Importance Of Land Policy For Long Term Development

In developing a long-term land policy, what can Zimbabwe learn from other parts of the world on this contentious land question? First, there are many countries in the world, particularly in Latin America and, of course, southern Africa where land is still a potentially explosive issue. As a generalisation, Asian countries have been more successful in land reform than African and Latin American countries and some scholars have offered opinions on this distinction. First is the observation that transforming agrarian systems into urban-industrial economies invariably requires fundamental changes in many institutions, including those of land tenure. The distribution of land ownership is a major factor that influences this transition from one form of social and political order to another. The experience of all industrialised and industrialising countries is the separation of a substantial segment of the ruling classes from direct ties to the land. Peter Dorner in his classic entitled *Latin American land reforms in theory and practice, a retrospective analysis* (1992) refers to the Asian experience in relation to Latin America. He cites the land reforms in Taiwan and South Korea and emerging economic powers, as having occurred early in their economic growth and industrialisation process, and that the industrial sector was never as closely tied to the in-egalitarian rural structures; as is often the case in Latin America. This observation is key to the future of Zimbabwe, South Africa, Namibia, and other African countries where land reforms are given a low priority, or where there is no political commitment for it.

Another common feature where land is a contentious political issue, and in particular where land distribution is highly inequitable, as is the case in southern Africa and Latin America, is the argument against land reform on the ideological basis that says that private property is a near-sacred right. Private property is then elevated to the status of foundation of a just and civilised society. But, as scholars have argued, if this promise holds', then it must likewise be accepted that private property cannot perform this noble function if most people are without it!

Land reform is a long-term process, not an event. Zimbabwe's land policy, therefore, should encompass some key values and principles, as well as a set of legal and administrative institutions that are effective and can deliver over a long period of time. Government over the last few years has developed a small but competent in-house capacity to implement such a vision. This paper is an attempt to seize the moment created by recent events on the land question. Various interested parties have developed positions on this issue.

Land Reform Has Been On The Minds Of Many Concerned Zimbabweans

In his own way and style, President Mugabe continues to show the courage to both Zimbabweans and the rest of the world, with a message that basically says that land re-distribution is a must, and secondly that this is a political objective he would like to accomplish for the eternity of history. Zimbabweans in general are in agreement and concerned about this issue, and recognise the need for lasting solutions that minimise short-term losses, be they economic or political. Since independence, and more so recently, various interest groups are active, and some loose associations have looked at this issue with varying levels of intensity.

This paper is an attempt to seize the momentum created by recent events on the land question. Various interested parties have developed positions on this issue. While this paper does not address directly these various contributions, it basically carves out a 'national strategy', largely based on government stated intentions, enriching that with positive elements out of the CFU's (Commercial Farmers' Union) contribution, as well as the various important points and issues raised at various stages of the public debate with the ZFU (Zimbabwe Farmers' Union), ICFU (Indigenous Commercial Farmers' Union) and other key stakeholders in the private sector and civil society. In early 1997 the authors of this paper convened a small group of concerned Zimbabweans who met to discuss the land issue. Then in July 1997, a bigger group met to discuss the desirability of a new strategy for land settlement in Zimbabwe that would meet short-term and long-term needs of the nation. Members of this informal group met several times before November 1997, after which their deliberations were overtaken by current events that culminated into the land identification exercise. The small group, however, was able to continue engaging a wider spectrum of views as consultations proceeded with government officials, farmer's unions, some private sector people and some donors.

Dimensions Of A National Strategy For Land And Agrarian Reform

A Vision Of The Major Goals

The Land Reform Programme will have three major strategic objectives. First is to achieve political stability, second is establishing a broader basis for economic growth, and third is the need for social integration. Land reform will explicitly target these objectives, and treat them as mutually inclusive and reinforcing. Land reform should address these three goals both in the short-term and long-term. Political stability will ultimately be reached through levelling the playing field in terms of land distribution and greater access to land by Zimbabweans through a vibrant land market where land becomes available at reasonable prices and reasonable and affordable sizes. Second is the growth of the economy and the availability of non-farm jobs at a rate greater than population growth rate. The combination of these two conditions will form the key ingredients of a Zimbabwean dream, where future Zimbabwean generations will have a reasonably fair chance to have a successful and fulfilling life, simply because such economic opportunities will be within grasp, irrespective of racial or economic family background. Economic growth and development is a key objective of this land and agrarian reform. Zimbabwe is still very much an agrarian society with the majority of Zimbabweans still rural and directly or indirectly relying on agriculture and the rural economy. It follows, therefore, that a land reform which by its very impact leads to a greater number of rural populace engaged in profitable and commercial farming, will unleash backward and forward rural economic growth linkages that will accelerate the growth of the whole economy. This is achieved as rural incomes rise and the demand for industrial goods rises. Food prices for urban workers will decline as food becomes more abundant. In addition, land markets will be integrated with rural financial markets leading to greater investment in both rural and urban industries. Social integration is an ultimate objective if Zimbabwe is to continue its growth into a peaceful and progressive non-racial society. Such social integration is desirable both for rural and urban industries. True integration is not forced, but rather, people and communities should be able to share common values across colour lines. Both economic development and political stability will contribute greatly towards the goal of social integration. Social integration is a major asset to society because cultural diversity in a situation of harmony enriches any society.

Principles and Values

A number of principles and values are drawn out from government's current land policy, as well as the experience of more than a decade of land reform, and these will guide the Land Reform Programme. The values and principles include, *inter alia*:

- greater security of tenure as an essential ingredient for commercial success;
- greater access to finance for land development and diversity in sources of finance for the programme so as to allow the mobilisation of a greater amount of financial resources;
- appropriate legislative and administrative provisions for land registration, lease management, farmer selection, and conflict resolution;
- greater efficiency of land use, more intensive farming, and divisibility of larger farms for greater access, greater economic utilisation, and a better distribution of land;
- a highly participatory and decentralised process that will empower and improve the civic participation of rural Zimbabweans in the land reform process;
- an effective farmer support system including financial backing, training and experience acquisition, as well as effective farm advisory and business management services.

The Urgent Need To Act

The single most serious enemy of land reform is not having a visible and meaningful practical programme that is running on the ground. The single most important task of government right now is therefore designing a national programme, with the backing of the key interested parties, that leads to immediate action. The courageous moves by the President Mugabe in gazetting land, and targeting 5 million hectares for compulsory acquisition, has already achieved the positive result of the CFU proposing the availing of 1.5 million hectares with immediate effect that will be willingly sold to government without the burden and delays of compulsory acquisition. It is of utmost urgency, therefore, that this be the basis of re-commencing the Land Reform Programme as this provides an avenue to kick-start the programme without major short-term political shocks and economic instability. Within a few years, greater capacity will be developed both financially and institutionally to implement the rest of the programme, and this phasing of the programme will most likely result in a win-win situation as the success of the programme brings greater commitment from all quarters. Progress towards this can be accelerated in the following ways:

- Speeding up land acquisition by Ministry of Lands and Agriculture as has already commenced with the CFU in identifying land that will be willingly offered to government without going the protracted compulsory land acquisition route. Ministry has to expeditiously assess this land for suitability for resettlement, and confirm the commitment of the farmers to availing this land;
- Government experts working in greater collaboration with experts from the private sector and all farmers' unions in preparing a comprehensive plan and fundable document that will be used immediately to source and mobilise funding for the programme. This should provide additional information and data for the government and budgetary process for the programme;
- Establishing a mechanism under government's auspices for continued high level consultation and joint planning with the private sector, farmers' unions, and other key players;
- Formulating a public relations and awareness programme that projects the national effort and growing consensus and commitment by all parties, as well as educating the media on the political and economic strengths of the programme.

Desired Elements Of The Proposed Land Reform Programme

Four new categories of land reform models are proposed for the future, and a number of new resettlement / reform options will be introduced under each so as to broaden access and flexibility. The four categories are:

1. Semi-commercialised small-scale settlement on newly acquired land;
2. Fully commercialised small to medium-scale settlement on newly acquired land;
3. Land tenure reform on existing resettlement areas;
4. Land tenure reform in communal areas.

Category 1: Semi-Commercialised Small-Scale Settlement on Newly Acquired Land

It is expected that the largest proportion of newly acquired land, may be at least 75 per cent, will be settled by small-scale land users from poor, landless and disadvantaged communities. Under this options 1 to 5 are offered as follows:

OPTION 1: Lease with option for title deed for small-scale mixed farms
It is proposed that most of the newly acquired land be settled under this option. Land acquired as large scale commercial land would be sub-divided and settled by selected farmers. Suitable candidates would be selected from Zimbabwe's rural population that includes the poor, the displaced, and the landless. Land would be settled under a 10-year lease with option to purchase and award of title deed. Settlers would generally not be expected to pay the market price for the land, but expected to pay a lease cost that covered the administrative cost. Land, however, would be expected to develop into fully commercial land within 10 years as owners developed and invested in this land and qualified for title deeds. This land would be planned and settled as stand-alone mixed farms for cropping and grazing.

OPTION 2: Long-term tradable leases for small-scale mixed farms
This is a variation of Option 1, with the provision that landowners who, within reasonable limits, are unable to develop the land, or unable to service the lease, will be allowed by the State to either convert to a 99-year lease, or trade the lease for value improvements to another land user who then has a full option for title. The option protects the poor against loss of rights in the event of lack of improvement, and the option still allows a commercial exit.

OPTION 3: Village settlement with family title for arable and residential land
Under this option an identified cohesive group of families would have option to settle on newly acquired land as a village, particularly on newly acquired land near communal land. In this case the families would be offered leases with option for family title deed for arable and residential land. The village assembly would acquire a village land title which would include the group title for common land. This option would be suitable for families and individuals who are part of a clan or neighbourhood moving from a communal area. It allows the semi-commercialisation of traditional

tenure, and reduces the cost and pain of resettlement, as well as provision of services and social infrastructure. The land affairs of the village would be locally managed by a village or primary court.

OPTION 4: Village CAMPFIRE settlement with title for land and rights over wildlife
This option would be appropriate where the existing land use is largely wildlife-based. Resettled families would be given an option to continue with wildlife-based land-use practice through a modified CAMPFIRE (Communal Areas Management Programme for Indigenous Resources) village settlement. These resettled families would be offered leases with an option for long leases or title deeds. Land users would maintain rights over wildlife in their area, and reap benefits through eco-tourism and other consumptive and non-consumptive commercial activities.

OPTION 5: Lease with option for specialised farming
Under this option, farmers would be settled for specialised production of high value crops such as tobacco, horticulture, dairy, tea, coffee, or sugar cane.

Category 2: Fully Commercialised Small to Medium-Scale Settlement on Newly Acquired Land

Under this category, which will account for say 25 per cent of the newly acquired land, indigenous commercial farmers would be settled on small-to medium-sized farms. All land would be settled on a lease with option for title deeds. Leases would generally be targeted for un-developed land. Where there are adequate improvements, then option for outright purchase and title should be offered. Settlers would be expected to pay the economic value of land and economic rates of interest (as opposed to market rates) over a period of up to 25 years. Acquired land would be sub-divided into small to medium-sized farms depending on intended land use and agro-ecological considerations. The following options are offered under this category as follows:

OPTION 6: Lease with title deed option for intensive small/medium scale mixed farm
This option would. be for traditional mixed livestock and cropping depending upon potential.

OPTION 7: Lease with title deed option for small / medium scale commercial tobacco farms

This option would be for those areas suitable for tobacco growing where large farms will be sub-divided into small / medium sized units as required for efficient and intensive tobacco production consistent with crop rotational needs and other environmental considerations. Settlers would be especially selected for this and effort put in place to avail them financially for training and technical advisory support.

OPTION 8: Lease with title deed option for small / medium scale commercial horticulture farms

This would be similar to option 7 but would target high value production and the value-added production of flowers, fruits, and vegetables, and for local and export markets.

Other specialised options are:

OPTION 9: Lease with title deed options for small / medium scale commercial dairy

OPTION 10: Lease with title deed for small-scale commercial wildlife / livestock farms

OPTION 11: Lease with title deed option for small-scale commercial sugarcane farms

Under fully commercial settlement, special provision would be made for rural residential land, and for registered companies.

OPTION 12: Rural residential settlement lease with title deed option

In keeping with the dreams of most Zimbabweans, individuals and families should have option for a residential plot on a planned rural suburb where there are prospects to build improved housing with water reticulation and lighting for the higher income groups, with provision for all income groups able to settle for residential land rights only.

OPTION 13: Registered companies for mixed and specialised farming on lease with title deed option

This replaces the previous 'Model B' of legal registered companies of commercial farming. Individuals would be offered leasing and title option.

Category 3: Land Tenure Reform on Existing Resettlement Areas

It is proposed that all existing resettlement areas should be offered option 14 or 15 immediately:

OPTION 14: Re-organised 'Model A' with individual farm units and lease with title option
Under this option the current permit system is replaced with a lease / title option after land is re-planned into individual farm units.

OPTION 15: Lease with title option for arable and residential land on 'Model A'
Under this option, the resettlement scheme would remain as it is, with arable and residential land, and common grazing. The current permit system would be replaced with a lease/title option for the arable and residential land, with a group title for common land.

Category 4: Land Tenure Reforms in Communal Areas

Option 16: Village title with option for Land Registration Certificates for arable and residential Land.
This would be as recommended by the Land Tenure Commission, 1993 (Rukuni, 1994).

Institutional and Legal Framework and Management of Programme

It is proposed that the institutional framework for this programme would be largely based on government infrastructure, and supported by civil society and other partners bringing various competencies. Current legal frameworks fall short of what would be necessary largely because of the absence of a clear and comprehensive land policy which deals with issues such as tenure security and support for land users. The formulation of such policy is currently being reviewed and is eagerly awaited.

Central Role of Government

It is proposed that current inter-governmental structures, led and co-ordinated by the Ministry of Lands and Agriculture and other key support ministries, will implement the programme. These structures would be supported by various inputs from other partners and key interested parties.

The National Land Board

To provide overall guidance and civic participation at higher levels, a National Land Board is proposed. The board could be established through an appropriate Act of Parliament, and report either through the Lands and Agriculture minister, or through the appropriate Minister of State in the Presidency. Eminent persons from across Zimbabwean society including government would be appointed to the board. Key interest groups would be represented, as well as other important sections of civil society. An independent chair would be appointed. The terms of reference of the board would be fleshed out by the Ministry of Lands and Agriculture so as to cover all key areas required in effectively overseeing and supporting the Land Reform Programme, and advising government on land policy. Such a board could play a vital role in enhancing the credibility of the programme, and providing additional 'think tank' support to government.

Public Relations and Awareness Campaigns

It is proposed that government would invest in a public relations and awareness campaign on the land issue. This task should be effective with the national plan being backed by the private sector and all the farmers' unions. It is important that Zimbabweans, as well as foreign nations, fully understand the legal, political and economic significance of the land reform programme, and the growing commitment and consensus for this programme and its progression.

Civil Society, Stakeholders and Provincial Land Boards

A decentralised participatory process would be ensured by the establishment of various bodies comprising various stakeholders and local communities, including the intended beneficiaries of the programme. These structures would be advisers to government at all levels, and particularly at provincial levels where Provincial Land Boards would provide the advisory mechanism for governors and governmental departments.

Financing the Land Reform Programme

Government and bilateral donors are the potential main source of funds with a concerted and aggressive programme being required to secure funds from these sources. This task is most urgent and requires a highly professional document to be prepared. Government would have to budget fully for the Land Reform Programme to incorporate funds expected from various sources. It is proposed that this task would be completed once the National Plan was in place. To date it has proven rather difficult to mobilise funds from financial markets and from donors. Commercial banks in particular have restrictions on long-term financing of land purchase. Donors, particularly multilateral, also face restrictions in financing land purchase and transfer. Both groups, however, desire to see a more vibrant land market, as well as a more supportive financial service for a growing and diversifying farming sector. In that regard, they are in a position to support other aspects of the programme besides land purchase.

Land Reform Trust Funds

In the light of the above, it is proposed that a Land Reform Trust Fund be set up with the following objectives:

- to mobilise funds initially from donors, philanthropists and other well-wishers, and ultimately from the local financial market;
- to channel those funds into various appropriate avenues including relevant government departments, as well as direct investment into commercial banks, agricultural banks and other appropriate financial institutions;
- to establish a long-term endowment fund for land settlement over the next 10-20 years;
- to provide policy guidelines for financial backing to newly settling farmers towards the purchase of land, as well as short to medium-term financing of farming operations;
- to establish a secretariat to co-ordinate the activities of the Trust.

It is proposed that the National Land Board may act as the board of trustees, or may appoint one to fulfil this role.

Conclusions

Zimbabwe has bravely swallowed some of the economic reform medicine over the last 8 years, and it is clear that the battle is not won until a number of fundamental and structural issues are addressed. Land reform is one of these, if not the most crucial. The political instability resulting from a malfunctioning economy and an unresolved and potentially explosive land reform issue are a real threat to future prosperity and social progress. It is therefore prudent to aim for a reform programme that addresses the three major objectives of political stability, economic development and social integration. There is scope and urgent need for a national plan backed by all key players including the private sector and civil society. An impenetrable plan has to be hatched within months, and collective responsibility is desirable in planning and financing the programme. This is achievable, and Zimbabwe's success with such an effort is needed in order to provide courage and commitment to other countries such as South Africa and Namibia, who still have to find lasting solutions to the land question 'time bomb'.

References

Boserup, E. (1981), *Population and Technological Change: A Study of Long Term Trends*, University of Chicago Press, Chicago.

Bruce, J.W., Migot-Adholla, S.E. and Atherton, J. (1993), 'The Findings and their Policy Implications: Institutional Adaptation and Replacement', in J.W. Bruce and S.E. Migot-Adholla (eds), *Searching for Land Tenure Security in Africa*, Kendall/Hung Publishing Company, Dubugue.

Dorner, P. (1992), *Latin American Land Reforms in Theory and Practice. A Retrospective Analysis*, The University of Wisconsin Press, Madison.

Feder, G. and Feeny, D. (1991), 'Land Tenure and Property Rights: theory and applications for development policy', *The World Bank Economic Review*, 5; 1, pp. 135-154.

Migot-Adholla, S.E., Hazel, P., Blorel, B. and Place, F. (1991), 'Indigenous land rights systems in Sub-Saharan Africa: A Constraint on Productivity', *The World Bank Economic Review*, 5, 1, pp. 155-175.

Moore, B. (1966), *Social origin of dictatorship and democracy*, Beacon Press, Boston.

Place, F., Roth, M. and Hazel, P. (1993), 'Land Tenure Security and Agricultural Performance in Africa: Overview of Research Methodology', in J.W. Bruce and S.E. Migot-Adholla (eds), *Searching for Land Tenure Security in Africa*, Kendall/Hung Publishing Company, Dubugue.

Roth, M., Barrows, R., Carter, M. and Kanel, D. (1989), 'Land Ownership Security and Farm Investment: Comment', *American Journal of Agricultural Economics*, 71: 211-14.

Rukuni, M. (1994), *Report of the Commission of Inquiry into Appropriate Agricultural Land Tenure Systems*, Government Printers, Harare.

Taylor, J. (1988), 'The Ethical Foundations of the Market', in V.Ostron, D. Feeny and H. Picht (eds.) *Rethinking Institutional Analysis and Development: Issues, Alternatives and Choices*, Institute for Contemporary Studies Press, San Fransisco.

Index

Postcript

Three months after the general election narrowly won by ZANU-PF, and generally judged to have been 'free but not fair'[1], the situation remains as confused as ever. Far from moving back to legality after winning, as many observers expected, the government seems to have set its sights on ensuring a victory for the ruling party in the 2002 election using the same tactics that have just won it a victory against the odds.

Far from concentrating on resettling the 804 farms surviving from the original 1,471 designated, government has now listed over 3,000 farms destined for takeover, covering about three-quarters of all commercial farms and half of the land area. It has also begun a "fast track" resettlement program that would move settlers on to many hundreds of the farms before the rainy season begins in November. More expected has been a widening division between the party and the 'veterans'. Some interpret this as the latter having served their purpose and now needing to be denied too much influence. There are also signs of a growing rift inside ZANU-PF, with some younger members questioning the sustainability of the current land reform programme, or indeed current economic policy generally.

Possible scenarios for the future range from government pressing on with taking over most of the commercial farms within a year as part of a professed new 'African radical' policy that would require a severing of many economic ties with the developed world, through to a fall of the government and victory for the MDC. This latter could presumably occur if the latter succeeded in winning enough of the nearly 30 seats where it has challenged the general election results. Short of this are a range of compromise possibilities for which a number of signs are already visible: the economy is in such dire straits that government has almost no room for manoeuvre and no resources to make land reform a success, let alone carry the (hopefully temporary) loss of production during reorganisation; rising unemployment and inflation again approaching 70%, with the economy expected to contract by 5% - 6% in 2000 are creating social unrest and divisions in the ruling party; and, in contradiction to its rhetoric in which the scope for land takeovers is widened, government has already ordered police to expel squatters (generally autonomous peasants rather than 'veterans') from some de facto resettled farms.

It is widely felt that President Mugabe's bid to stay in power has not only already cost the country several years of growth, but also seriously set back the prospects for viable and sustainable land reform, rather than advanced it. It does seem inevitable that the current situation will force a major political crisis in Zimbabwe or some significant political compromise.

Whatever the political make-up that results, with the country in such dire economic circumstances, the country will inevitably be beholden to International Financial Institutions. The extent to which, and the way in which, programmes of economic recovery can also accommodate land reform may well have to be renegotiated.

It is hoped that the many messages of this book can help inform the rebuilding of a land reform programme that is both viable and sustainable, allowing all Zimbabweans once again to work together for the prosperity of the country.

Note

1   The conduct of the election itself, involving few serious breaches, allowed the ANC majority on the South African parliamentary observer group and the SADC and OAU observer groups to maintain that the election was broadly free and fair "by African standards" and enough African members of the Commonwealth Observer Group agreed so as to ensure that the group's report was anodyne. However, according to the report of the Helen Suzman Foundation: "The events which occurred in the four months prior to the election were in our opinion, such as to preclude the possibility of a genuinely free and fair election - and, indeed, this was exactly what they were intended to do. It is not acceptable to say that the election was broadly free and fair "by African standards" for this is to accept the racist view that normal democratic standards are somehow inappropriate for Africans." Many NGOs including Amnesty International shared this view.

Printed and bound by CPI Group (UK) Ltd, Croydon, CR0 4YY

22/10/2024

01777605-0005